T0358962

CORROSION CONTROL FOR OFFSHORE STRUCTURES

CORROSION CONTROL FOR OFFSHORE STRUCTURES

Cathodic Protection and
High-Efficiency Coating

First edition

By

RAMESH SINGH

AMSTERDAM • BOSTON • HEIDELBERG • LONDON
NEW YORK • OXFORD • PARIS • SAN DIEGO
SAN FRANCISCO • SINGAPORE • SYDNEY • TOKYO
Gulf Professional Publishing is an imprint of Elsevier

Gulf Professional Publishing is an imprint of Elsevier
225 Wyman Street, Waltham, MA 02451, USA
The Boulevard, Langford Lane, Kidlington, Oxford, OX5 1GB, UK

First edition **2014**

Notice
No responsibility is assumed by the publisher for any injury and/or damage to persons or
property as a matter of products liability, negligence or otherwise, or from any use or
operation of any methods, products, instructions or ideas contained in the material herein.
Because of rapid advances in the medical sciences, in particular, independent verification of
diagnoses and drug dosages should be made.

Library of Congress Cataloging-in-Publication Data
A catalog record for this book is available from the Library of Congress

British Library Cataloguing in Publication Data
A catalogue record for this book is available from the British Library

For information on all Gulf Professional publications
visit our web site at store.elsevier.com

This book has been manufactured using Print On Demand technology. Each copy is produced
to order and is limited to black ink. The online version of this book will show color figures
where appropriate.

ISBN: 978-0-12-404615-3

Working together
to grow libraries in
developing countries

www.elsevier.com • www.bookaid.org

DEDICATION

This book is dedicated to the memory of my beloved Rui who left us before I could complete the book.

Now that the book is completed and I am free to take her on walk, I do not have Rui by my side.

CONTENTS

ACKNOWLEDGMENT

Writing this book made me realize the complexities of the subject. In normal work life we often tend to think in a relatively unidirectional way about a subject and we focus on the task ahead.

In trying to explain the same subject, the topics that we take for granted need to be explained in detail. I do not pretend to have explained every one of them, but I have tried to cover as many as I could.

A book like this can never be said to be all inclusive for two very specific reasons.

1. There are so many supporting subjects, all complex enough to be a book in itself; and
2. The changing nature of challenges, which are further complicated by the improving technologies.

I am extremely grateful to my employers and clients who gave me opportunities to learn new technologies and face new challenges in the field. How could I have learned without these opportunities?

In writing this book, I also thank my colleagues who always encouraged me to take up this project. I am also indebted to the encouragement, support, and help from my friend Olga Ostrovsky. She helped me negotiate the obstacles of writing and editing the drafts. Without her expert help this book would not be possible.

Last but not least, I am also grateful to my loving wife Mithilesh, and my son Sitanshu for their support in accomplishing this goal. Mithilesh tolerated me while I concentrated on the project. Without her support and understanding this task was not possible.

ABOUT THE AUTHOR

Ramesh Singh, MS, IEng, MWeldI, is registered as Incorporated Engineer with British Engineering Council, UK, and is a Member of The Welding Institute, UK. He is a Senior Principal Engineer (Materials, Welding, and Corrosion) for Gulf Interstate Engineering, Houston, TX, USA (rsingh@gie.com; Tel.: +1-713-850-3687). Ramesh Singh is a graduate from Indian Air force Technical Academy, with diplomas in Structural Fabrication Engineering and Welding Technology. He has been a member and an officer of Canadian Standard Association and NACE and served on several technical committees. He has worked in industries spanning over aeronautical, alloy steel castings, fabrication, machining, welding engineering, petrochemical, and oil and gas industries. He has written several technical papers and published articles in leading industrial magazines, addressing the practical aspects of welding, construction, and corrosion issues relating to structures, equipment, and pipeline.

He provides consultancy services on management of welding, materials, and corrosion expertise through Nadoi Management, Inc. (nadoi51@yahoo.com; Tel.: +1-832-724-5473).

PREFACE

Materials on this subject can be found in various international specification and project files stored in the archives of engineering companies, and I can confidently say that NACE International (www.nace.org) has greatly advanced the subject over the years.

Other bodies that that have played prominent roles in developing and promoting the technology are Det Norske Veritas (DNV) (www.dnv.com) and the International Organization for Standardization (ISO) (www.iso.org), and many publications address corrosion, coating, and cathodic protection.

In this book, I intend to introduce to the concept of cathodic protection. With the foundation of knowledge provided by the book, the practitioner can explore and reference the industry data and specifications and then learn the subject by practicing it. This book can also be used as guide for cross-referencing and validating a practitioner's current understanding of corrosion prevention.

For readers trying to steer through the subject, this book provides useful guide for balancing the theory and practice of cathodic prevention. My objective is to keep budding engineers moored in the theory of offshore structure corrosion control, as well as its applications. The book can also serve as a reference for engineers, nonengineers, managers, and inspectors.

For readers seeking more detailed studies on any specific topic covered in this book, specialized associations and institutions can offer further guidance, and there are several published works available from these bodies that can support the reader in developing an in-depth understanding of specific subjects related to corrosion control.

Basics Concept of Corrosion

CHAPTER ONE

Need for the Study of Corrosion

INTRODUCTION

From the perspective of environment, the corrosion activity is a natural phenomenon that is necessary for sustaining the natural balance. Corrosion is a great leveler of engineering materials in that it tries to revert the metal back to its most stable form. However, from an engineer's perspective, corrosion could be seen as a destructive attack of nature on metal. This destruction of metal is, however, brought about by nature's chemical or electrochemical reaction. It causes significant loss of material which leads to losses in terms of productivity and cost of maintenance, repair and replacements, and restoration. This does not included damage to property and the occasional loss of lives and injuries associated with failures resulting from corrosion.

Corrosion occurs in every aspect of modern life; every life is affected, though not everyone is aware of the corrosion happening around them. It is an all pervasive and 24/7 activity.

The impact of corrosion is three directional, the three aspects being economic, safety, and environment. The impact of corrosion, and the prevention thereof, is felt economically, and affects the safety and environmental conservation of resources. In the succeeding discussions we will see how these three aspects manifest themselves.

- Economically, it implies the loss of infrastructure by way of loss of materials used in tanks, process equipment, pipelines, platforms, bridges, and many other important structures. The economic losses could be direct or indirect. The direct losses would include, for example, the cost of replacing the corroded structures, equipment, and the cost of painting, upkeep, and monitoring of cathodic protection as well as the associated labor cost. Another cost would be the use of expensive corrosion resistance materials.
- The indirect cost of corrosion is difficult to assess accurately as more complex aspects come into play. However, activities that can be counted

as contributing to the indirect cost of corrosion might include the closing of plants and facilities for repair and maintenance needed because of corrosion damages and failures. These costs add up because shut down involves reduction in production, loss of product, costs for cleaning and repair of environmental damages, and wages paid for the duration of the nonproductive time. In a nutshell, it can be said that indirect losses are a chain of activities that will take place and have to be paid for even when production is not there to support those costs.

- The loss of structure materials to corrosion is not only an economic loss but it makes the structures weak and degrades their designed capabilities and reduces the structure's designed purpose. On the extreme end of this deterioration, such structures can become a safety hazard and the loss may lead to structure failures, some of which could even be catastrophic, leading to property damage and loss of lives.

- The metal resources are primarily extracted from naturally occurring oxides of metal called ore, and if corrosion is allowed to degrade the metals, more and more resources will be required, leading to more environmental damage.

- Materials, especially metals, are required to sustain the infrastructures for the development of civilization; however, care must be taken to reduce the impact of the growing demands of civilization on the environment. This should be one of the primary responsibilities of engineering. The balancing of nature and development is handled by responsible engineering and the process is a vicious cycle because the damage to the environment may itself threaten the very civilization for which the extractions are made. The human civilization has reached such a stage that, while corrosion itself is a natural recycling process, the prevention of corrosion leads to the reduction of damage to the environment.

The economic impact is possibly the prime motivator for the study of corrosion and the development of preventive measures by the industry. The other two aspects discussed above are, to a great extent, taken care of in the process. The reality of the situation is that a compromise between the development costs and the environmental impact should be sought. Whether it is adequate or not is, and will always be, a point of contention between environmentalists and development planners. The engineers can provide the best possible middle ground by studying corrosion and suggesting preventive steps. Thus, engineers have a great responsibility to balance the needs of the developing society and sustaining the environment.

The cost of corrosion in the United States is estimated to be about 4.3% of gross national product according to one study done in 1978. It is estimated that about one-third of that cost can be saved by the application of corrosion control measures. As per a current estimate, in the developed countries the cost of corrosion is estimated to be between 3.5% and 4.5% of the GNP. There are various ways that engineers can design the process and equipment with an aim to reducing this cost.

A brief look at some of the losses that are directly related to failures from corrosion would justify the need for a better understanding of corrosion.

COST OF CORROSION

We have highlighted the cost of corrosion in terms of GDP; we now expand this in more realistic terms as we start with the impact of corrosion on environment.

- Cost of environmental damages

  No economic value can justify the actual loss to the environment caused by the failure of a structure. This is not including the cost of cleaning and the regulatory fines and actions that would directly affect the finances of a company.

- Production loss and down time due to corrosion damage

  This requires that the process of corrosion is stopped and repairs are carried out. This leads to reduction in production output and reduced revenues, which, when added to the cost of standby labor and fuel losses, becomes a major cost. This, however, does not include the loss of market and public image.

- Accidents

  Corrosion failures are, by themselves, accidents. However, what is meant by the term accident here is the loss of lives and injuries caused by failures. Corrosion damages can cause severe accidents resulting in injuries and loss of lives. It adds to the economic loss, too. It also projects adverse public image and loss of market share.

- Product contaminations

  For many industries, the contamination of product with corrosion can lead to severe quality issues and loss of business reputation, and sometimes may even result in serious implications to people's lives and health.

- Loss of efficiency; over designing leads to excessive cost of energy

In some cases, corrosion is anticipated and portions of a system are designed to address the expected corrosion. However, lack of a full understanding of the corrosion and its impact can lead to over designing of the equipment and system which often leads to an inefficient system, which, in turn, uses excessive energy to operate and maintain the plant, adding this cost to the final product.

• Increased capital cost

Over designing also increases the capital cost of the project; this may include a poorly designed corrosion monitoring and mitigation system. If all these are not convincing reasons to acquire proper knowledge of corrosion then what else could justify the importance of its study?

The study of corrosion involves nearly all engineering faculties, including metallurgical, mechanical, aeronautical, chemical, electrical, and civil to name just few, and the application of some pure sciences such as physics, chemistry, biology, and, of course, mathematics and statistics. So, although this study often appears to incline to metallurgical engineering, where it ultimately rests, it is equally encompassing of all other disciplines of study. This implies that all engineering faculties should have a good program to educate on the primary understanding of the essentials of corrosion.

We have very briefly discussed the impact of corrosion and the need for the study of corrosion but, up to this point, we have not defined the term. We need to know what corrosion is, and what conditions are necessary for corrosion to occur. For that we need to find a definition for corrosion. We also need to know how to understand various types of corrosion and how they can be identified, measured, and mitigated.

A keen reader might have noted that we have not associated the word eliminate with corrosion because that is simply impractical if not impossible. This is because, as we said in the beginning, corrosion is a natural phenomenon, it cannot be eliminated. Recognizing the fact that corrosion is a natural phenomenon, we have used words like "control" and "mitigate" in speaking of the impact of corrosion. The effect of corrosion and the damages caused by corrosion in a given environment can only be contained, it cannot be eliminated.

In subsequent chapters, various methods of corrosion control, monitoring, and mitigation practices will be discussed, with emphasis on corrosion control and mitigation of offshore structures and pipelines of various descriptions.

CHAPTER TWO

Corrosion Principles and Types of Corrosion

In the previous chapter, we introduced the need for the study of corrosion, but we did not define corrosion. Though nearly everyone sees some very visible examples of corrosion in their daily lives, mostly in the form of brown rust deposited on steel and iron surfaces, there are several other forms of corrosion that are not recognized by the average person. This is because not everyone knows what corrosion is or how it manifests itself in different environments and on different materials. This introduction obviously leads us to the question; what is corrosion?

Because the corrosion phenomena is vast, there can be several responses to this question and that can lead to different definitions of corrosion. While they all may address a specific type of corrosion, none of them encompass all aspects and forms of corrosion. Thus, in an attempt to define corrosion we also discuss the principles that affect corrosion. One NACE publication defines corrosion as "The deterioration of a substance—usually a metal—or its properties because of reaction with its environment."

From this NACE definition we understand that corrosion is the interaction between a material and the environment to which it is exposed. This leads us to accept the fact that understanding corrosion involves understanding both the material and its environment. Environments that cause corrosion could be atmospheric, the presence of certain liquids, high-temperatures, or being underground—which is similar to being in certain liquids. The other aspect of corrosion is the material; the corroding material could be metallic or nonmetallic; in our current scope, we focus on corrosion of metals. To understand corrosion it is essential that the metallurgical aspect of the material is fully understood, including the fundamentals of the structures and properties of engineering materials, metals, and nonmetals. To understand corrosion of a metal it is important to know the path taken to refine that metal from its natural form (ore) to its usable form, and subsequent processing and any heat treatment that may have been applied to the material to make it useful.

In common speech as well as in the NACE definition above, the word corrosion signifies deterioration of material, an element of natural degradation is implied. There is an element of correctness with this common understanding—there is degradation and nature is taking part in that degradation.

The above definition is a good description of the physical appearance of the action of corrosion. However, it does not address every aspect of the scientific explanation of corrosion. For example, it does not explain the flow of energy and thermodynamic reactions involved in the corrosion process. Although the thermodynamic reaction could be grouped with the environment aspect of the corrosion, it is, in itself, a very prominent aspect that needs independent study.

The following discussion complements the discussion we have had so far and addresses additional scientific explanations to bring about a better understanding of the term corrosion, one that is suitable for engineers.

FLOW OF ENERGY

Corrosion occurs in nearly all materials produced by nature or manmade materials; this includes metals and nonmetals such as certain plastics, ceramics, and concrete. In order to find the correct definition of the word corrosion, a question arises, why do metals corrode? A search for the answer leads us to the realm of thermodynamics.

Thermodynamics is the science of the flow of energy. This science explains a specific corrosion process and indicates if corrosion is possible in a given metal and environment. The flow of energy in the corrosion process is in the form of electrical energy. The rate of corrosion is similarly predicted by the kinetics. We discuss these topics in more detail in later sections of this book. As we know, with the exception of a few naturally occurring metals, most engineering materials are found in the form of ores, often metal oxides found in nature. A lot of energy is spent in the extraction process of these usable metals from their ores. Hematite (Fe_2O_3) is an ore of iron, and bauxite ($Al_2O_3 \cdot H_2O$) is an ore of aluminum, there are some more complex ore like that of Nickel ore kupfer-nickel, smaltite ores are a combination of sulphur and arsenic which are roasted to form an oxide which is then reduced to the metal by hydrogen and purified by the Mond process to obtain nickel that engineers can use. Copper is found as pure metal; that is the reason copper is usually free from corrosion, but it is also extracted from various ores like Ruby ore (Cu_2O), Copper Glance (Cu_2S), or Pyrite ($CuFeS_2$). It may be pointed out that copper obtained from nature and copper extracted from ores would display different potentials. As we can see, a lot of energy is put into the extraction of engineering metals. Some metals that are extracted as free metals from the earth and do not require additional energy to convert the natural form to make it usable are very low in corrosion galvanic energy. These metals are called Nobel metals. In the galvanic table, Table 2.1, metals are listed from the more negative potentials (Active) to more positive (Noble).

In the galvanic series table, a new term, "potential," is introduced in relation to corrosion—it is one way of measuring the energy difference between two metals. Electrons flow from a higher energy state anode, which is a negative, to a low energy electrode, a cathode. The potential difference between two electrodes facilitates the flow of electrons. If a voltmeter of sufficient sensitivity is attached across the flow circuit, the potential difference between anode and cathode can be measured. The potential of each metal in reference to another is a unique number, these numbers by themselves do not establish a standard by which to measure and compare all possible potential differences and thus it is not of much practical use except in relation to the two metals. One unified scale is needed to compare and establish a universal reference for understanding the potential difference of various metals. For this purpose, a standard electrode is used; a reference electrode is so constructed that its potential is reproducible. There are a number of standard electrodes that are used as reference electrodes. Some of the common

Table 2.1 Galvanic Series

Active (more Negative Potential) End	Metals
↕	Magnesium
	Zinc
	Aluminum alloys
	Carbon steel
	Cast iron
	13Cr (Type 410) Steel (Active)
	18-8 (Type 304) Stainless Steel (Active)
	Naval brass
	Yellow brass
	Copper
	70-30 CuNi alloy
	13Cr (Type 410) Steel (Passive)
	Titanium
	18-8 (Type 304) stainless steel (Passive)
	Graphite
	Gold
Noble (more positive potential) end	Platinum

Table 2.2 Reference Electrodes

Reference electrodes	Potential (V)
Calomel (0.1 M)	+0.3337
Calomel (1.0 M)	+0.2800
Copper–copper sulfate	+0.3160
Silver–silver chloride (dry in sea water)	+0.25
Calomel (saturated)	+0.2415
Silver–silver chloride (saturated)	+0.2250
Silver–silver chloride (molar)	+0.2222
Hydrogen	0.0000

reference electrodes include Colomel, Copper–Copper Sulfate, and Silver–Silver Chloride and these, along with their potentials are listed in Table 2.2.

For most common engineering materials corrosion is electrochemical in nature and occurs in an aqueous environment. The aqueous environment is the electrolyte in the corrosion process, through which the electron travels from anode to cathode (i.e., high-potential to low-potential metal). The corrosion process involves the removal of electrons (oxidation) of metal and the consumption of those electrons is termed reduction reactions, often indicated by the presence of oxygen or reduction of water from the aqueous environment, electrolyte, etc. The following reactions illustrate these points.

$$Fe \rightarrow Fe^{++} + 2e^- \text{ (removal of electrons } - \text{oxidation process,}$$
$$\text{an anodic reaction)}$$

$$O_2 + 2H_2O + 4e^- \rightarrow 4OH^- \text{ (presence of oxygen } - \text{reduction and evolution}$$
$$\text{of hydrogen ions, a cathodic reaction)}$$

$$2H_2O + 2e^- \rightarrow H_2 + 2OH^- \text{ (aqueous environment } - \text{reduction and}$$
$$\text{evolution of hydrogen ions of water, a cathodic reaction)}$$

The oxidation reaction is an anodic reaction in which the metal loss occurs, while the reduction reaction is a cathodic reaction. Both these reactions are electrochemical in nature and are essential for corrosion to occur. The oxidation loss at the anode must be balanced with consumption of emitted electrons at the cathodic end. This charge neutrality is very essential for the corrosion process to occur and continue. If, however, this is not maintained and the accumulation of large negative charges between the metal and electrolyte occurs, then gradually the reaction reduces and eventually stops the corrosion process. This is utilized to an advantage in a cathodic protection system as we discuss in later chapters, where the process of providing Cathodic Protection is used to prevent metals from corrosion.

Oxidation and reduction reactions are referred to by various names (e.g., Red-ox and Half-Cell reaction). These reactions can occur within a metal itself (Half-Cell) even though they are not physically separated; when they are physically separated the reaction is referred to as a corrosion cell.

From the above basic description of corrosion, we can deduce that a corrosion cell must have the following four components to be active.

1. An anode (the location from where electrons are emitted and metal loss occurs).
2. A cathode (the location where electrons collect).
3. A metallic path (often the structure itself provides that metallic path).
4. An electrolyte, in which the anode and cathode are immersed (the electrolyte could be any moist surface or immersion in any conducting fluid, water, or soil).

If we remove any one of these four essential elements from the corrosion cell, the corrosion action will stop.

The electrochemical aspect of corrosion gives us an opportunity to detect and mitigate the corrosion of structures. We can monitor the potential difference measured in volts and the currents associated with the existing corrosion process.

MEASURING THE DIFFERENCE OF ENERGY LEVELS FOR DETECTION OF CORROSION CELLS

The energy difference between anode and cathode can help us determine if there exists a corrosion cell. This is, in fact, the measure of electrical potential between two points. The energy difference causes a potential difference which causes the flow of electrons. If a voltmeter is inserted in the circuit, we can measure the potential difference in volts.

ELECTROCHEMICAL MECHANISM OF CORROSION

As stated, the corrosion process is mostly an electrochemical process. In a corrosion reaction, a partial electrochemical step occurs which is influenced by the electrical variables that include the current flowing (I), potential difference (voltage V), electrolyte to metal interface, and the extent of electrical resistance to electrical current flow.

In an aqueous media, the action is similar to a dry cell, where the carbon electrode in the center and the zinc cap is separated by sodium chloride (Na_4Cl) electrolyte. In this cell the carbon electrode in the positive pole, at the carbon electrode chemical reduction occurs and at negative pole is the zinc electrode where oxidation takes place. In this reaction, metallic zinc is changed into zinc ions, thus corroding the zinc electrode. The rate of corrosion is linked to the rate of electricity produced. This relationship is quantitative and it is explained with the help of Faraday's law.

$$\text{Weight of metal reacting} = klt$$

where k is a constant for the metal, for zinc this value is 3.39×10^{-4} g/C (gram per coulomb); time is t in seconds; and l is the flow of current in amperes.

One coulomb is the measure of electricity produced by the flow of 1 amp, for 1 s.

Table 2.3 has a list of electrochemical equivalents of various engineering materials.

In the dry cell example, the metal zinc is in an active cell, the metal is constantly corroding; however, if the circuit is broken and the current stops flowing, the zinc stops corroding. Any other metal surface in a similar situation behaves similarly, that is, the metal surface corrodes if it is in the electrical circuit and stops corroding if removed from that electrical circuit. If the metal is embedded with impurities, the current can flow from one spot to

Table 2.3 Electrochemical Equivalency of Various Engineering Metals

Element	Symbol	Valency	Electrochemical equivalent
Vanadium	V	5	0.10560
Nickel	Ni	2	0.30409
Iron Fe 1 0.57865	Fe	1	0.57865
Cadmium	Cd	2	0.58244
Platinum	Pt	4	0.50578
Vanadium	V	5	0.10560
Chromium	Cr	3	0.17965
Lead	Pb	2	1.07363
Mercury	Hg	1	2.07886
Gold	Au	1	2.04352
Tin	Sn	2	0.61503
Tungsten	W	6	0.31765
Tantalum	Ta	5	0.37488
Silver	Ag	1	1.11793
Copper	Cu	1	0.65876
Zinc	Zn	2	0.33876
Lithium	Li	1	0.07192
Aluminum	Al	3	0.09316

another within that metal as an electrical cell is created within the metal itself. This phenomenon of electrical circuits in the presence of an electrolyte is called local action cell or simply local cell. Any process that reduces the presence of impurities in a metal improves the corrosion resistance of that metal. This explains why purified aluminum and magnesium are more resistant to corrosion in seawater than several of their commercial versions. But this does not mean that pure metals do not corrode.

As we have briefly pointed out, the local action cell causes corrosion, but this local cell can be caused not only by inherent impurities but also by the variations of structure within the metals, temperature, and environment. The best examples of these are the corrosion cells between the parent metal and weld metal, the corrosion of high-purity iron in air saturated with water which is nearly the same as commercial grades of iron in water.

GALVANIC CELL, ANODE, AND CATHODE

Let us take the example of a cell discussed earlier. When two electrical conductors called electrodes are immersed in an electrolyte, they create a galvanic cell. A galvanic cell converts chemical energy into electrical energy. If these electrodes are connected by a wire, current flows through the metallic conductor. This flow is from the positive electrode to the negative electrode. The direction of flow follows an arbitrary convention; however, in actuality, the electrons go from negative to positive poles.

Within the electrolyte, the current is carried by both negative and positive carriers; these carriers are one or more atoms that are electrically charged and are called ions. The current carried by ions is dependent on the mobility of the electrical charge, but the total of positive and negative current in the electrolyte of a cell is always equal to the total current carried by the metallic conductor. Ohm's law describes this as

$$I = E/R$$

where I is the current in amperes, R is the resistance of the conductor in ohms, and E is the potential difference measured in volts.

The positive electrode is where the reduction takes place; where positive current enters the electrode through the electrolyte is called the cathode. Chemical reduction reactions are also called cathodic reactions. Examples of reduction reactions, or cathodic reactions, are given here:

$$H^+ \rightarrow \frac{1}{2}H_2 - e^-$$

$$Cu^{2+} \rightarrow Cu - 2e^-$$
$$Fe^{3+} \rightarrow Fe^{2+} - e^-$$

Conversely, the electrode where the oxidation occurs is called the anode; at this point the positive electricity leaves the electrode and enters the electrolyte. The following examples are chemical reactions that represent anodic reaction.

$$Zn \rightarrow Zn^{2+} + 2e^-$$
$$Al \rightarrow Al^{3+} + 3e^-$$
$$Fe^2 \rightarrow Fe^{3+} + e^-$$

Often the anode is the point where the corrosion occurs. Exceptions are the alkaline reaction products that form at the cathode and sometimes cause secondary corrosion of amphoteric metals like aluminum, tin, lead, and zinc. These metals corrode rapidly if exposed to acid or alkalis.

Here we take a little diversion from the discussion to explain the behavior of metals in the corrosion process.

Metals have plural phases; the alloying elements determine the types and number of phases the specific metal will have. The atoms in metal donate part of their outer electrons to the gas electrons, the gas electrons are from the gas that has diffused into the metal. This exchange of atoms causes electrical conductivity which is established to be 10^5 S/cm. However, pure elements which do not react electrochemically as a single component do not have such constant conductivity value. The value is approximated and a median value is assumed; for iron, which is the main element in our discussion, the approximation can be understood through the following reaction, in which both components can react with the electrolyte.

$$Fe \leftrightarrow Fe^{2++} = 2e^-$$

In the first reaction, metal dissolved by the passing of Fe^{2+} results in appositive current (I_A) and a metal loss (Δm) occurs. This is an anodic reaction. The second reaction is a cathodic reaction in which the passing of the electrons results in negative current I_C, but no metal is lost. These two reactions are reversed in the transfer from solution to the metal. In the anodic reaction, electrons are transferred to the metal, whereas in the cathodic reaction, the metal is deposited. Using the principle of the second reaction, the cathode metal deposit is used in electroplating, and, as is evident, it is the reverse of the corrosion process.

In both reactions, Faraday's law applies: metal loss (Δm) is a function of the metal's atomic weight (M) and referred electrical charge (Q) on one hand and valance of metal ions (z) and Faraday's constant (F) as denominators. This relationship can be expressed through the following equation.

$$\Delta m = MQ/zF$$

The above relationship can also be expressed in terms of the specific gravity of the metal, where the current density $I_{density}$ is a function of the atomic weight of the metal (Q) and is inversely related to the function of time (t) and the surface area of the electrode, as is expressed in the following equation.

$$I_{density} = Q/St$$

Keep in mind that corrosion parameters are dependent on time and place and must be modified to suit specific time- and place-related reaction velocities. Different local removal rates are basically caused by the varying compositions and nonuniform surface films. This variance results in different thermodynamic reactions and kinetic effects which also influence this reaction.

• The uniform weight loss, also called general corrosion, can occur on active surfaces of single-phase metals.

There are no passive films on the active metal surfaces.

The local corrosion cells are discussed further later in the book; however, it may be added here that:

• The selective corrosion is only possible in multiphase alloys.
• Shallow pitting in general is only possible in the presence of surface films, and if they are damaged, particularly on passive metals like austenitic steels, high chromium steels Ni alloys, and aluminum and Titanium and their alloys and other metals with passive surface films.

This brings us to the subject of metal removal and the passing of electrons through the electrolyte. The metal removal is not a direct reaction of this passage of electrons. It is an indirect activity that is related to the law of electron neutrality, as expressed in the following relationship.

$$I_A = I_C$$

The above relationship is limited by the fact that there are conditions where electrons may be diverted thorough a conductor.

Electrons cannot be dissolved in an aqueous solution; however, they react with oxidants in the solution.

$$Ox^{n+} + ze^- = Red^{(n-z)+}$$

In the equation, the Ox is the symbol of oxidation and the Red is the reduction media. The numerical charges of oxidation and reduction are represented by n and $n-z$ symbols. When the electrochemical redox reaction as indicated is absent, the corrosion rate Δm as described is zero as a result of the electron neutrality discussed earlier. The metals that have passive films, where the films act as insulators, behave in a similar pattern.

The oxidizing agents concentrate, and this concentration is essential for the reaction in the above reduction oxidation equation. The reaction can be further divided into oxygen corrosion and acid corrosion. The oxygen reaction is represented by the following reaction:

$$O_2 + 2H_2O + 4e^- = 4OH^-$$

The acid reaction is represented by the following:

$$2H^+ + 2e^- = 2H \rightarrow H_2$$

The evolution of hydrogen from the acid molecules can occur in slightly dissolved weak acids like H_2CO_3 and H_2S and hydrogen can also be produced by the acid molecules. In this case, the concentration of acid produces more aggressive corrodents than the measure of the pH value. As the following reaction suggests, the hydrogen can also evolve from the water molecules.

$$2H_2O + 2e^- = 2OH^- + 2H \rightarrow 2OH^- + H_2$$

The above reaction occurs with cathodic polarization currents. If, in the natural solution, the conditions support free corrosion of steel, then this reaction can be safely ignored. In some cases, if the amount of gas evolved or consumed is in excess of the reactions discussed, then it is essential that the volume of evolved gas is determined.

All metals can experience electrolytic corrosion without contribution of external currents. The fractures caused by cathodic hydrogen can occur only when the absorbed hydrogen tensile stress reaches a critical value. For hydrogen to be absorbed some promoters are necessary. But there are exceptions to this, very low pH and very negative potential hydrogen absorption can occur. Steels with hardness in excess of 350-HV are particularly susceptible. More on this aspect when we discuss stress corrosion cracking and sulfide stress corrosion cracking.

As shown in the equations in the earlier section, generally, corrosion can be classified as an electrochemical process and subsequent control of these equations can prevent corrosion from occurring. Corrosion occurs at the anode as this is where electrons are released.

Certain metals, especially the metals in subgroup 4 and 5 (see Table 2.4) of the periodic table are especially prone to hydrogen embrittlement and fracture as they form internal hydrides. The materials listed in the main group, 4-6 form volatile hydrides and they are prone to weight loss corrosion. The type of corrosion arising from cathodic hydrogen can be the limiting factor in the application of cathodic protection and this must be kept in mind when designing a cathodic protection system. In a galvanic cell, the cathode is the positive pole and the anode is the negative pole.

When current is impressed on a cell from a battery, rectifier, or generator, the reduction occurs at the electrode connected to the negative pole of the external current source, this electrode is the cathode. The electrode connected to the positive pole is the anode, as in the electroplating

Table 2.4 Periodic Table

Group	1	2	3	4	5	6	7	8	9	10	11	12	13	14	15	16	17	18
Period																		
1	1 H 1.008																	2 He 4.0026
2	3 Li 6.94	4 Be 9.0122											5 B 10.81	6 C 12.011	7 N 14.007	8 O 15.999	9 F 18.998	10 Ne 20.180
3	11 Na 22.990	12 Mg 24.305											13 Al 26.982	14 Si 28.085	15 P 30.974	16 S 32.06	17 Cl 35.45	18 Ar 39.948
4	19 K 39.098	20 Ca 40.078	21 Sc 44.956	22 Ti 47.867	23 V 50.942	24 Cr 51.996	25 Mn 54.938	26 Fe 55.845	27 Co 58.933	28 Ni 58.693	29 Cu 63.546	30 Zn 65.38	31 Ga 69.723	32 Ge 72.63	33 As 74.922	34 Se 78.96	35 Br 79.904	36 Kr 83.798
5	37 Rb 85.468	38 Sr 87.62	39 Y 88.906	40 Zr 91.224	41 Nb 92.906	42 Mo 95.96	43 Tc [97.91]	44 Ru 101.07	45 Rh 102.91	46 Pd 106.42	47 Ag 107.87	48 Cd 112.41	49 In 114.82	50 Sn 118.71	51 Sb 121.76	52 Te 127.60	53 I 126.90	54 Xe 131.29
6	55 Cs 132.91	56 Ba 137.33	* 71 Lu 174.97	72 Hf 178.49	73 Ta 180.95	74 W 183.84	75 Re 186.21	76 Os 190.23	77 Ir 192.22	78 Pt 195.08	79 Au 196.97	80 Hg 200.59	81 Tl 204.38	82 Pb 207.2	83 Bi 208.98	84 Po [208.98]	85 At [209.99]	86 Rn [222.02]
7	87 Fr [223.02]	88 Ra [226.03]	** 103 Lr [262.11]	104 Rf [265.12]	105 Db [268.13]	106 Sg [271.13]	107 Bh [270]	108 Hs [277.15]	109 Mt [276.15]	110 Ds [281.16]	111 Rg [280.16]	112 Cn [285.17]	113 Uut [284.18]	114 Fl [289.19]	115 Uup [288.19]	116 Lv [293]	117 Uus [294]	118 Uuo [294]

*Lanthanoids	*	57 La 138.91	58 Ce 140.12	59 Pr 140.91	60 Nd 144.24	61 Pm [144.91]	62 Sm 150.36	63 Eu 151.96	64 Gd 157.25	65 Tb 158.93	66 Dy 162.50	67 Ho 164.93	68 Er 167.26	69 Tm 168.93	70 Yb 173.05
**Actinoids	**	89 Ac [227.03]	90 Th 232.04	91 Pa 231.04	92 U 238.03	93 Np [237.05]	94 Pu [244.06]	95 Am [243.06]	96 Cm [247.07]	97 Bk [247.07]	98 Cf [251.08]	99 Es [252.08]	100 Fm [257.10]	101 Md [258.10]	102 No [259.10]

process. Due to this confusing description of the negative and positive concept, it is advised that the concept of positive and negative be dropped and you should memorize and reference only in terms of cathode, where current *enters from electrolyte*, and anode, where current *leaves to return to the electrolyte*.

The corrosion of metals, in particular steel in an aqueous environment which can be either soil or water, occurs because the metal interacts with the local environment. In the case of steel, man has mined iron ore and processed it into steel. However, due to certain characteristics of the steel, it is not "stable" once in contact with an aqueous environment and interacts with the local setting in an attempt to return to its naturally occurring state. This process is corrosion.

The basic process at an anodic site is the release of iron (Fe) from the steel surface into the environment and this can be expressed as the following reaction.

$$Fe \rightarrow Fe^{2+} + 2e^-$$

During the process, two electrons $(2e^-)$ are generated which must be consumed by the environment (in aerated systems) and can be expressed as:

$$4H^+ + O + 4e \rightarrow 2H_2O$$

A summary of these half reactions can be expressed as:

$$2Fe + 2H_2O + O_2 \rightarrow 2Fe(OH)_2$$

$Fe(OH)_2$, iron oxide, can be oxidized to form the red-brown $Fe(OH)_3$ commonly referred to as rust. The corrosion three stages of cells are shown in Figure 2.1.

CATIONS AND ANIONS

When electricity flows through a cell, the ions that move toward the cathode are called cations. These are always positively charged ions. In the similar electrical flow through a cell, the negatively charged ions are called anions. Anions and cations exist in water; they carry the current through the migration of the electrical field. When such conditions exist, the electrolyte is termed an aqueous electrolyte.

The electrical conductivity is of interest in the corrosion process in cell formation (e.g., as we see when discussing types of cells). Conductivity is

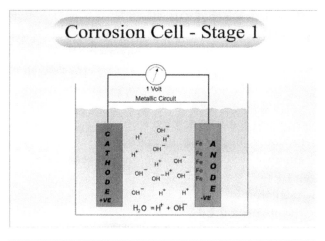

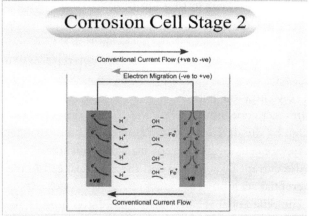

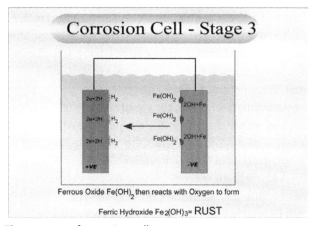

Figure 2.1 Three stages of corrosion cells.

increased in the salt-dissolved electrolytes. It is important to note that salts may not be part of the corrosion itself but they increase the conductivity and thus provide a more corrosion-aggressive environment. The corrosion rate of carbon steel (of which most offshore structures are made) in brine is not affected by dissolved salts in the presence of oxygen, as discussed earlier in the reduction oxidation reaction.

The same is not true for the localized corrosion as the dissolved salts have very strong reactions. The presence of chloride ions that accumulate at local anodes can stimulate dissolution of iron and prevent the formation of a protective film.

The corrosion process identifies three types of cells. These are the active cells that are responsible for corrosion of metals; they may coexist or may be active independently. We briefly introduce these cells as we reference them frequently in our discussions in later chapters.

1. Concentration Cells

   These cells have two identical electrodes and each is in contact with a solution of differing composition. There are two types of concentration cells.

   a. Salt concentration cells

      In a salt concentration cell, the current flows because the cell is created by the difference in the concentration of the electrolyte. In this type of cell, the process equalizes the concentration of salts caus- ing the current flow to stop and the electrical cell to die.

   b. Differential aeration cells

      The differential aeration cell is more common in the corrosion field, especially in crevice corrosion and pitting corrosion. This cell develops because of the different level of air (oxygen) in the electrolyte. The difference in oxygen level due to the variation of aeration causes the current flow. However, the polarity may reverse depending on the aeration level, stirring rate, or if the two metals are short circuited.

2. Differential Temperature Cells

   In differential temperature cells, the electrodes are of the same metals and in the same electrolyte; however, the temperature of the electrode varies. This initiates a corrosion cell. The best examples of such cells can be described as corrosion in boilers, heat exchangers, and similar equipment where the potential difference is created by the variation in metal (elec- trode) temperature placed in the same electrolyte.

   In a cell in which two iron electrodes, one relatively hotter than the other, are immersed in dilute aerated sodium chloride solutions, the hot electrode is anodic to the colder metal of the same composition. This type of cell is normally unstable. In this type of cell, the polarity can reverse if the temperature gradient changes or the cell may die if the two temperatures equalize.

3. Dissimilar Electrode Cells

The best example of this type of cell is the simple dry cell discussed earlier. Two electrodes of dissimilar metal are immersed in the same electrolyte. Dissimilar metals, such as bronze and steel, in same electrolyte are another good example of this type of cell. Materials of the same grade in different work conditions can have a potential difference and can form these cells. Even a single metal containing impurities can form its own local cell in an electrolyte. Various crystal phases of the metal tend to show different potentials, leading to corrosion within, but the differing potentials of crystals often achieve equilibrium, and corrosion stops. This is because the most corrodible planes of an atom react first, as compared with the least corrodible planes, resulting in the least corrodible planes being the only faces exposed, regardless of the original orientations.

In this type of cell, the corrosion rate continues to differ because of the differing surface areas of differing crystal faces. The least corrodible crystal face of any metal is not always the same but varies with the environment.

For example, the study by R. Glauner and R. Gloker in *Z. Kristalloger* published in 1931 points out that the 110 face of copper corrodes most rapidly in 0.3N HCL-0.1N H_2O_2, but in 0.3N HNO_3-0.1N H_2O, the 111 and 110 faces are most reactive.

A similar study by H.M. Howe, published in *Metallography of Steel and Cast Iron*, McGraw-Hill, New York, 1916, points out that, in dilute Nitric acid, the 100 face of iron is least reactive.

TYPES OF CORROSION

In the earlier discussion, we referenced general metal loss as general corrosion and localized corrosion. There are various types of corrosion, some are very common and can be seen in day-to-day life, while there others are rarely seen except in very specific combinations of material and environments. The study of corrosion involves such detailed knowledge of specific forms and their specific environments.

To better understand corrosion, and then to find the right mitigation and control methods, we need to know the different environments for various types of corrosion.

1. General Corrosion

General corrosion is commonly recognized as "rusting" of iron, or fogging of nickel, or tarnishing of silver. The rate of uniform attack is reported in various units: the most accepted terminology in penetration of corrosion is mil per year, millimeter per year (mm/year), or inch per year (ipy), and in terms of weight loss measure, it is reported in grams per square meter per day (gmd) (a mil is a thousandth of an inch [0.001 in] and is equivalent to 0.0254 mm).

Examples of time-averaged general corrosion rate values for steel corrosion in a seawater environment are reported as 0.13 mm/year, (mm per year) or 0.005 ipy (inch per year). R.W. Revie and N.D. Greene, published in *Corrosion Science 9775*, 1969, conclude that the corrosion rate is greater in the initial stages; hence, the duration of exposure time should always be reported. Extrapolating the data is not always safe in this regard.

Metal density also plays a significant role in the corrosion rate; hence, knowledge of metal density is essential for correct reporting of corrosion rate in ipy or mm/year. The given weight loss per unit area for light metal signifies a much greater actual loss than the same weight loss of a heavy metal.

For general corrosion purposes, the rate of corrosion is classified variously by different writers, but NACE in its corrosion data survey reports four levels, as follows:

- Less than 2 mils (0.05 mm)/year
- Less than 20 mils (0.508 mm)/year
- 20-50 mils (0.508-1.27 mm)/year
- Greater than 50 mils (1.27 mm)/year

Uhlig and Revie, in *Corrosion and Corrosion Control*, describe the rate of corrosion in three stages, as follows:

- <0.15 mm/year (0.005 ipy or 5.9 mil): Metals in this category have a good corrosion resistance to the extent that they are suitable for critical parts (e.g., valve seats, pumps shafts, and impellors).
- 0.15-1.5 mm/year (0.005-0.05 ipy or 5.9-59 mil): Metals in this category have a satisfactory performance, if a higher rate of corrosion can be tolerated (e.g., tanks, piping, valve bodies, and bolt heads).

- > 1.5 mm/year (>0.05 ipy or >59 mil): Metals in this category are usually not satisfactory.

Common applied techniques for corrosion prevention and control are based on the general corrosion rate. The principle of corrosion prevention is to provide an electrical barrier between the cathode and anode (e.g., the coating and passivation of the surface).

Changing direction of the flow of current through the structure into the electrolyte is another approach that covers cathodic and in some cases anodic corrosion protection methods.

In most situations, especially on carbon and low-alloy application, a combination of an electrical barrier between cathode and anode, and flow of current control method are used as an effective corrosion control method. We discuss both of these methods in more detail in Sections 3 and 4.

2. Localized Corrosion

General corrosion is the uniform spread of corrosion over a large surface area in the form of rust or tarnish. In contrast, localized corrosion occurs on specific sites on the metal surface and in specific environmental conditions. Corrosion activities on these specific sites may start and stop with a change of the environment. The areas surrounding the corrosion sites are often unaffected by corrosion.

In localized corrosion, the rate of corrosion on the same material is greater in some areas than others.

Several different types of localized corrosion that are very specific to the environment and metal include:

- Pitting
- Fretting
- Cavitation
- Crevice
- Filiform

These different types of localized corrosions and what conditions promote them are:

- Pitting

  Pitting is a deep narrow corrosive attack which often causes rapid penetration of the substrate. Pitting is localized as the surrounding metal remains unaffected. The initial (the start of the pit) activity starts with a "flaw" in the metal surface that has different potential than the surrounding metal surface. A flaw could be caused by a breakdown of passive layers, such as: in stainless steels, coating holidays, change in the metals structure, or microbiological reactions caused by bacteria, just to name few.

Once the mechanism has started, the continuum mechanism of pitting involves a cell between the interior of the pit and the external surface. The interior of the pit often contains corrosive acids and hydrolyzed salts. An anode establishes within the pit and the surrounding surface becomes the cathode, initiating a corrosion cell between the interior of the pit and the external surface.

Deep pitting is a deep depth of the pit concentrated in a small area of metal that is acting as anode (e.g., the corrosion of a stainless steel immersed in seawater shows deep pitting). Conversely, shallow pitting is a shallow depth of the pit spread over a larger area (e.g., the corrosion of an iron pipe buried in soil shows shallow pitting).

Pitting factor is the ratio of deepest metal penetration to an average metal penetration determined by weight loss of the specimen.

The pitting corrosion occurs in stages, these stages can be identified as:

- *Initiation*: Pitting initiates at the defect sites or imperfections in protective coating or at the local loss of passive films as in stainless steels. Once the initiation begins the corrosion activity is rapid.
- *Propagation*: Corrosion is driven by the potential difference between the anodic pit and the external surface of the metal surrounding the pit. Often the environment within the pit acts like an electrolyte and the cathodic area outside at the edge of the pit, completing all the four elements required for sustaining a corrosion cell.
- *Termination*: A pit may terminate if the internal of the pit has stopped being an anode (e.g., by filling the pit with corrosion product itself or by the removal of the corrosive environment).
- *Reinitiating*: The pit can restart if the terminated pit is rewetted or corrosion product is removed.
- Fretting

Fretting is a localized corrosion phenomenon that is caused by *mechanical* stresses. It is the result of slight relative motion (for example, as in vibration) of two substances of which at least one of them is a metal, that are in contact and were not intended to have relative movement in that fashion. Fretting manifests itself as a series of pits at the metal surface. Metal oxide debris usually fills the pits masking the pits, unless the corrosion product is removed.

Such damage occurs at the interface between two surfaces that can move with respect to each other. The surfaces are free of any corrosion products and appear burnished. This may result in oxidation of the surfaces or galling, seizing, or fatigue, eventually causing the cracking of the materials involved.

- Crevice

Crevice corrosion is caused by the potential difference in the concentration of materials inside and outside the crevice, often as the result of a design flaw that restricts the free flow of the environment between the surfaces.

Crevice corrosion is a form of corrosion in which the site of the corrosion has restricted access to the surrounding environment. The crevices occur at the interface of metal and metal or metal and nonmetal.

The mechanism of crevice corrosion can be due to either oxygen concentration cells or metal ion concentration cells.

The oxygen concentration cell is caused by cathodic potentials with respect to areas that are anodic. The most common reaction of this type is the reduction of oxygen or water in one of the following modes, in which oxygen is the reactant.

$$2H_2O = O_2 + 4e^- \rightarrow 4(OH^-) \text{ Water reduction}$$
$$O_2 + 4H^+ + 4e^- \rightarrow 2H_2O \text{ Oxygen reduction}$$

The basic principle is that an increase in the concentration of reactants will further drive the reaction., implying, in turn, that a build-up of reaction product will stifle the reaction. In other words, the buildup of corrosion product will stifle the corrosion process in the crevice.

- Filiform

Filiform is caused by the specific oxygen cell occurring beneath the organic or metallic coating on materials. It appears as a fine network of random threads of corrosion product developed beneath the coating material. Under relative humidity in excess of 60%, filiform often appears on surfaces beneath the coatings.

The corrosion mechanism for filiform is similar to that of crevice corrosion, as this is driven by the potential difference between the advancing head of attack and the area surrounding it. This form of corrosion advances in a filament of about 0.1 mm (0.004 in.) wide, where the head of the filament becomes anode, as compared to the trailing cathodic area. The pH of the area immediately behind the head of the filament is very low, also with notable lack of oxygen. The corrosion follows the cathode as it chases the anode head. Metal oxides are formed as a corrosion byproduct.

- Cavitation
  Cavitation erosion results from the formation and collapse of vapor bubbles at a dynamic metal–liquid interface. This collapse of bubbles causes pits, and often cavitation appears as a honeycomb of relatively deep fissures. An example of this type of corrosion can be found on the rotors of a pump or on the trailing edges of ships' propellers.

3. Galvanic corrosion

This form of corrosion occurs due to the potential difference between two metals or between metals and conductive nonmetals when they are electrically connected in an electrolyte environment. In effect, this configuration of materials forms an electrochemical cell. The electrical connection may be direct or through any external conductive path.

The impact of the nonmetals in galvanic corrosion is demonstrated by the conductivity of nonmetals, such as the presence of carbon and graphite in plastics and corrosion products, such as magnetite (Fe_3O_4) and sulfides, that are often cathodic with respect to most metals.

Ions of more noble metals may be reduced on the surface of a more reactive metal, and the resulting deposit acts as a cathodic site, potentially causing more galvanic corrosion cells on the surface of the more reactive metal.

Galvanic corrosion can be recognized by either of the two conditions.

- Increased corrosion of anodic material
- Decreased corrosion of the cathodic material

The corrosion effect is more pronounced where dissimilar materials are placed immediately adjacent to each other in a corrosive environment.

As stated earlier, galvanic corrosion is an electrochemical cell reaction. The presence of all four elements listed below is necessary for galvanic corrosion to occur.

- An anode
- A cathode
- A metallic-path
- An electrolyte

The electrons flow through a metallic path from anodic sites to cathodic sites in the presence of an electrolyte, which allows the ions to flow.

In most galvanic corrosion sites, all four essential elements are readily available. The metallic path, anode, and cathode are present in the metals

Table 2.5 Hydrogen Solubility

Alloy	Crystalline Lattice	Solubility: cm³ of Hydrogen/100 g of Metal
Alpha iron	bcc	0.7
13% Cr-Bal Fe	bcc	0.4
13% Cr-Bal Fe	fcc	4.8
18% Cr-10%Ni Bal Fe	fcc	5.8

themselves, and the only externally required element is the electrolyte, which is not so difficult to find. Atmospheric moisture could fill in that requirement.

Table 2.5 presents a list of materials and their galvanic potentials in seawater.

In understanding galvanic corrosion, it is important to know the electrochemical reaction. The Nernst equation establishes that reaction. This equation relates to the potential of pure metals in solutions containing various concentrations of ions. The potential of a pure metal is a ratio of reaction products to reactants. The Nernst equation describes this ratio as follows. relationship.

$$E = E° - RT/nF(\ln)$$

In the above equation, E is the actual reaction potential, $E°$ is the potential under standard conditions, (all activities $= 1$), R is the gas constant, T is the temperature in degrees Kelvin, n is the number of electrons transmitted in the reaction, F is the Faraday's constant (which is taken as 96,500 Coul/Mol), and ln is the natural logarithm.

Activity $= 1$ for metals in their metallic state and 1 for ions in 1 M concentration (1 mol/w/L), which is roughly equal to concentration of ions in terms of molar concentration in dilute solutions.

Assuming the standard temperature of 22 °C (about 72 °F) replacing the actual values of constants R, T, and F, and converting from the natural logarithm to normal base 10 logarithms, the Nernst equation can be rewritten as following:

$$E = E° - 0.059/n(\log)$$

The Nernst equation applies to many corrosion reactions and to the potentials of reference electrodes that we will discuss in detail when we consider cathodic protection in subsequent parts of this book.

Galvanic corrosion affects the anode in the couple by increasing its corrosion rate. This corrosive attack may manifest itself as a general attack, a localized attack, such as pitting, or some other form of corrosion.

Galvanic corrosion activity does not control the form of corrosion that occurs at the anode; it only increases the rate of attack.

The principle of galvanic corrosion is put to best use in the protection of metallic surfaces by making the vulnerable surface a cathode by introducing another metal in the circuit that is anodic to the material to be protected. We will discuss this concept as cathodic protection by sacrificial anodes in subsequent sections and chapters of this book.

AREA EFFECT IN GALVANIC CORROSION

Galvanic corrosion also depends on the area and the geometric and distance effects related to it.

The area effects in galvanic corrosion relate to the ratio of cathodic surface area to anodic surface area. In the immersion conditions, an anode that is small with respect to the cathode will corrode faster than a large anode. If the anode area is increased so that it is lager than the cathode area, the corrosion of anode (i.e., depletions of anode) will then be made negligible, however.

In atmospheric conditions (nonimmersed), the affected area is only the wetted contact area, normally a very small area, and the effective area ratio is near to 1:1.

In immersed conditions, most of the corrosion occurs at the junction of the couple. This is affected by the resistivity of the electrolyte. In high resistivity electrolytes, the effect at the junction between the anode and cathode is more pronounced than in low–resistivity electrolyte. A cell's geometry also affects the extent of current flow. Galvanic current tends to concentrate at sharp points and edges. More about the shape of the anode is discussed in Sections 2 and 3 of the book, in which anode manufacturing, testing, and current output in cathodic protection systems are described.

4. Environmental Corrosion Cracking
   Environmental corrosion cracking is the brittle failure of otherwise ductile material. This type of cracking is caused by the combined action of corrosion and tensile stress. Other forms of corrosion happen over the time, thus increasing the chances that the corrosion will be detected

during routine inspections. Yet, environmental corrosion often outpaces inspection and detection, leading to a quick spread of the damage and subsequent failure.

Environmental corrosion manifests itself as tight cracks normal to the applied tensile stress. Corrosion products are found imbedded in the cracks that may appear as multiple cracks in fine network on the surface.

The involved material, its tensile stress, the environment, and the temperature are the contributing factors in this form of corrosion, and logically, these factors are also control the process.

There are several types of recognized environmental corrosions that we shall introduce and briefly discuss.

• Hydrogen-induced cracking

Hydrogen-induced cracking (HIC) is a brittle failure of an otherwise ductile material. HIC is the combined effect of tensile stress and hydrogen in the metal. Atomic hydrogen, also referred as nascent hydrogen, is produced on the metal surface, and it can be absorbed by that metal, leading to cracking. High-strength alloys with tensile strength of 150 ksi or greater are more susceptible to HIC.

The assessment of tensile stress must also account for the residual stress caused by fabrication and other combined stresses. As stated, HIC is a brittle failure of an otherwise ductile material, when exposed to an environment where hydrogen can enter through the metal's surface. This is a cathodic phenomenon where the normal evolution of hydrogen at the cathodic site is inhibited, and atomic hydrogen in the cathodic reaction enters into the metal. Thus, the atomic hydrogen in the presence of poisons, such as cyanides, arsenide, antimonides, phosphides, and sulfides, is inhibited from forming molecular hydrogen. The atomic hydrogen defuses into intestacies of the metal, instead of evolving into a gas molecule as a result of a cathodic product reaction.

The above description demands more explanation of hydrogen and its effect on metals. Certain conditions have to be present for hydrogen to enter a metal and be absorbed through its surface. As stated above, nascent hydrogen can enter the metal surface, while molecular hydrogen cannot. Nascent hydrogen is very potent and is produced on metal surfaces as a result of the decomposition of process gases such as ammonia or hydrogen cracking, or as a corrosion product or during welding, heat treatment, pickling, or electroplating.

Atomic hydrogen can defuse through metal lattices and remain there because of its smaller size as compared with molecular hydrogen. Within the steel, the behavior of hydrogen is a function of its solubility and its diffusivity in steel.

The solubility of hydrogen in steel is a function of following factors.

- The phase of steel: if solid or liquid
- The crystalline structure of steel
- The alloy content of steel
- The temperature of steel in solid condition
- The partial pressure of hydrogen in the environment

In pure iron, the solubility of hydrogen at room temperature is very limited; however, as the temperature is raised, the solubility increases. At the temperature where alpha iron (bcc lattice) transits to gamma iron (fcc lattice), which is at about 1670 °F (910 °C), the hydrogen solubility is significantly high. At this temperature, it is estimated that the solubility is about 3 cm^3/100 g of iron.

As the temperature is progressively raised, we further note that there is a pattern in the increasing solubility of hydrogen in steel.

When gamma iron (fcc lattice) is further heated to a point where gamma iron transforms to delta iron (bcc lattice), the solubility of hydrogen takes another leap. In this temperature range, the solubility is about 5 cm^3/100 g of iron.

As the heating continues, a large increase is noted at the melting point and beyond, and in this range, the solubility is reported to be between 13 and 27 cm^3/100 g of iron. The variation relates to the measurement of hydrogen at partially molten or fully molten stages of iron.

Alloying elements also influence the absorption of hydrogen in steels, as alloys create different crystal lattices, and crystal lattices have significant influence on hydrogen solubility. As we have noted above, at a given temperature, the fcc lattice can contain a much larger amount of hydrogen than the bcc lattice can. Alloying elements such as chromium decrease the solubility of hydrogen. On the other hand, nickel up to 8% in solution can only slightly increase solubility of hydrogen in steel.

Table 2.5 gives a glimpse of some alloys, lattice structures, and hydrogen solubilities.

The influence of hydrogen partial pressure on hydrogen solubility follows the relations established by Sievert's law given below.

According to this law, the solubility is proportional to the square root of the partial pressure of hydrogen in the system:

$$S = k \times p^{0.5}$$

Sievert's law is effective to a certain pressure level only. It does not apply when the pressure exceeds 12 Mpa or about 1760 psi, and temperatures exceeding 200 °C (about 400 °F), where methane could be generated within the steel, due to the hydrogen attack. This temperature limit is for carbon and low-alloy steel, and the temperature limit for chromium and molybdenum steels is significantly higher.

Cold work also influences the solubility of hydrogen. Cold work exceeding 15% reduction in area increases the solubility of hydrogen, and hydrogen trapped in cold work lattices is difficult to remove by bakeout.

Steel exposed to a gaseous hydrogen environment is more susceptible to hydrogen–related failures. Steel that transforms from gamma to martensite with trapped hydrogen is more susceptible to cracking upon welding.

Sulfide stress cracking (SSC) is a specific type of HIC in which the presence of sulfide inhibits the formation of molecular hydrogen. NACE MR 0175, which is also ISO specification number with ISO 15156, provides details on how to determine conditions that may lead to SSC, while offering guidance on the selection and qualification of material suitable for SSC.

- Corrosion fatigue

As the name implies, corrosion fatigue is caused by the combined effect of fatigue stress and a corrosive environment. A metal that progressively cracks on being alternately or repeatedly stressed is said to fail by fatigue. The greater the applied stress during each cycle, the shorter the time to failure. An S-N curve of stress versus number of cycles can be plotted for any material that is subjected to stresses caused by fatigue. Such a curve also depicts the stress under which no fracture occurs, and this stress value indicates the metal's endurance limit.

For steels the endurance limit can be safely assumed to be half of the tensile strength. The terms fatigue strength and endurance limit are often used interchangeably to describe this value. For other metals, this relationship is not entirely true. Hence, for other metals and complex alloy steels, it is safe to say that the stress below which fractures do not occur is the metal's fatigue strength. For such metals and alloy steels, an S–N curve should be developed by plotting the results of a number of tests.

The above brief discussion of fatigue stress and S–N curves relates to steel that is not immersed in a significantly corrosive environment. In a corrosive environment, failure at a given stress level usually happens within very few cycles, thus occurring much faster than the process represented by the original S–N curve. As a result, a true fatigue strength in corrosive environment is no longer observed. The cracking of metal due to the combined effect of a corrosive environment and alternating stress is called fatigue corrosion.

As described above, it is a premature failure of a cyclically loaded member in a corrosive environment. This failure often occurs at significantly lower stress, or at fewer numbers of cycles, than it would in a noncorrosive environment.

The cracks resulting from fatigue corrosion are typically transgranular (tin and lead showing the known exceptions), and they appear branched, with several smaller cracks accompanying a major crack on the metal's surface. Cracks often form at the base of the corrosion pit; however, a corrosion pit is not a necessary element for fatigue corrosion cracking. One of the best known and widely discussed forms of corrosion fatigue is the stress corrosion cracking (SCC).

• Stress Corrosion Cracking

There are two types of SCC normally found on pipelines;

(1) The high pH (pH 9-13) and

(2) Near-neutral-pH SCC (pH 5-7).

The high pH SCC caused numerous failures in the United States in the early 1960s and 1970s, whereas near–neutral–pH SCC failures were recorded in Canada during the mid–1980s to early–1990s. The SCC failures have been reported from all over the world, including Australia, Russia, Saudi Arabia, and South America.

SCC defined

SCC is a brittle failure mode in otherwise ductile material. This unexpected and sudden failure occurs in ductile metal that is also under tensile stress in a corrosive environment. This condition is further aggravated if the metal is at elevated temperature in a pipeline, which, for this purpose, is a temperature above 40 °C (104 °F).

SCC is an anodic process; this is verified by the application of cathodic protection, which is sometimes used as a remedial measure. Usually, there is some incubation period prior to the cracks being detected, and during this period, the cracks remain at a microscopic level. This is followed by the active progression of the cracks. Sometimes cracks may be self-arresting, as in multibranched transgranular SCC, because of localized mechanical relief of stresses.

If the stress or the specific environment responsible for SCC is removed, the SCC will not occur, and further progression of cracks will stop. The SCC is associated with little general corrosion. In fact, if extensive general corrosion is present, SCC is less likely to occur. However, trapped corrosion product may initiate another SCC cell under the general corrosion.

There is general understanding that steel that has tensile strength above 130 ksi (896 MPa) is susceptible to SCC, but this does not mean that steels below that level of tensile strength are not susceptible to SCC. As is said above, the stress and the specific environment are the main contributor to the cause of SCC, and temperatures above 40 °C (104 °F) add to that condition.

Factors Essential to Cause SCC

Environment

SCC is highly chemical specific, and certain materials are likely to undergo SCC only when exposed to a small number of chemical environments. The specific environment is of crucial importance, and only very small concentrations of certain highly active chemicals are needed to initiate SCC. Often the chemical environments that cause SCC for a given material are only mildly corrosive to the material in other circumstances. As a result of this phenomenon, the parts of a structure with severe SCC may appear unaffected on casual inspection, despite being filled with microscopic cracks. Unless a conscious effort is made via a specific targeted inspection plan to detect SCC, this special condition of SCC can mask the presence of SCC—cracks—for a long time. SCC often progresses rapidly leading to catastrophic failures.

The second part of SCC is the stresses. Stresses can be the result of the crevice loads due to stress concentration, or they can be caused by the type of assembly or residual stresses from fabrication (e.g., cold working). The residual stresses caused by fabrication can be relieved by annealing.

Environment is a critical causal factor in SCC. High-pH SCC failures of underground pipelines have occurred in a wide variety of soils, covering a range in color, texture, and pH. No single characteristic has been found to be common to a large number of soil samples examined in such cases. Similarly, the compositions of the water extracts from the soils have not shown any more consistency than the physical descriptions of the soils. On several occasions, small quantities of electrolytes have been obtained from beneath disbonded coatings near locations where stress corrosion cracks were detected. The principle components of the electrolytes were carbonate and bicarbonate ions, and it is now recognized that a concentrated carbonate-bicarbonate environment is responsible for this form of cracking. Much of this early research focused on the anions present in the soils and electrolytes. In addition, the coating failure, the local soil, temperature, water availability, and bacterial activity have a critical impact on SCC susceptibility. Coating types, such as coal tar, asphalt, and polyethylene tapes, have demonstrated susceptibility to SCC. High efficiency coating systems, such as three-layer polyethylene (3LPE) and Fusion bonded epoxy, have not shown susceptibility to SCC.

Loading

Loading is the second most important parameter contributing to SCC, especially cyclic loading. The crack tip strain rate defines the extent of corrosion or hydrogen ingress into the material. There has been no systematic effect of yield strength on SCC susceptibility.

Other factors

Certain types of welds, especially low-frequency welded ERW pipe, have been found to be systematically susceptible to SCC. Nonmetallic inclusions have also had limited correlation to SCC initiation.

High-pH SCC

High-pH SCC can be called classical SCC. The phenomenon was initially noted in gas transmission pipelines. In the practical terms, it is often found within 20 km (about 12.5 mile) downstream of a compressor station. High-pH SCC normally occurs in a relatively narrow cathodic potential range (-600 to -750 mV Cu/CuSO$_4$) in the presence of a carbonate and bicarbonate environment in a window from pH 9 to pH 13. The system temperature should also be greater than 40 °C (104 °F) for high-pH SCC to occur. The crack growth rates decrease exponentially with decreasing temperature.

An intergranular cracking mode generally represents high-pH SCC. A thin oxide layer is formed in the concentrated carbonate-bicarbonate environment, which surrounds the crack surfaces and provides protection. However, due to changes in loading or cyclic loading, there is crack tip strain resulting in breakage of the oxide film. This results in crack extension due to corrosion. Because of such a stringent environmental requirement for high-pH SCC initiation, this type of SCC is not as prevalent as the near-neutral-pH environment SCC. The high-pH SCC has been primarily noted in gas transmission lines associated with higher (greater than 40 °C) temperature.

High-pH SCC Mitigation Strategy

To evaluate and establish the extent of SCC susceptibility, the following steps must be taken and considered.

a. Evaluate the selection of material, coating, and other operational conditions that are conducive for SCC.

b. Conduct an over-the-ditch coatings survey to identify locations of holidays and match them with high stress levels. A high stress level is defined as stresses equal to or exceeding 60% of the specified minimum yield strength of the involved material.

c. Identify and match the stress with high-temperature locations.

d. Match the inspection report and identification of coating failures with corrosion, even minor corrosion, to identify the potential for SCC.

e. Excavate to identify susceptibility, conduct magnetic particle inspection on suspected locations to locate SCC, and meet mandatory requirements and due diligence inspection.

Near-Neutral-pH SCC

The near-neutral-pH SCC is a transgranular cracking mode. The phenomenon was initially identified in Alberta, Canada, and it has also been noted in reports from pipeline operators in the United States. The primary environment responsible for near-neutral-pH SCC is diluted groundwater containing dissolved CO_2 gas. As in high-pH SCC, the CO_2 is generated from the decay of organic matter. Cracking is further exacerbated by the presence of sulfate-reducing bacteria (SRB). This occurs primarily at the sites of disbonded coatings, which shield the cathodic current reaching to the pipe surface. This creates a free corrosion condition underneath the coating, resulting in an environment with a pH around 5-7.

A cyclical load is critical for crack initiation and growth. There are field data that indicate that, with a decreasing stress ratio, there is an increased propensity for cracking. Hydrogen is considered a key player in this SCC mechanism, as it reduces the cohesive strength at the crack tip. Attempts have been made to relate soil and drainage type with SCC susceptibility; however, limited correlations have been established.

There has been no correlation to a clear threshold for SCC initiation or growth. The morphology of the cracks is wide, with evidence of substantial corrosion on the crack side of pipe wall.

Near-neutral-pH SCC Mitigation Strategy

To evaluate and establish the extent of SCC susceptibility, the following steps are taken and evaluated.

    a. Evaluate the material selection and coating system to ensure they are compatible with the SCC conditions.

    b. Review and analyze the corrosion inspection survey reports to identify areas of corrosion, as well as linear or small pitting corrosion locations to identify sites for SCC susceptibility.

    c. Identify and analyze locations of high cyclical pressure combined with a high operating pressure.

    d. Conduct bell-hole inspection and excavate at several of these locations to develop extent of SCC on the pipeline system.

    e. Conduct magnetic particle inspection to identify the presence off cracks.

    Additional parameters, such as soil and drainage, should also be considered for SCC susceptibility; however, the pitfalls of this step must be born in mind, as both very poor and well-drained soils have shown susceptibility to SCC.

- Liquid Metal Embrittlement

This embrittlement manifests itself as decreased ductility when it is in contact with liquid metal. The failure is brittle at low stress levels.

This is not a time-dependent failure, and cracking may occur quickly on application of stress.

5. Flow-Assisted Corrosion
   Flow-assisted corrosion occurs due to the combined action of corrosion and fluid flow.
   This can be divided in three types:

   a. Erosion corrosion
   b. Impingement
   c. Cavitation
      • Erosion corrosion
        Erosion corrosion occurs when the velocity of the liquid is sufficient to remove protective films from the metal surface. The corrosion sites are often localized in areas where flow aberrations and turbulences occur due to material discontinuities. Such discontinuities could be the weld, any notch in the metal, or the very geometry of the metal itself.
        In flowing liquid or gas, if the velocity is sufficient to remove the corrosion product from the surface, thus reducing the protective effect, it may further remove the exposed substrate, causing an accelerated loss of metal by way of corrosion and leading to its failure.
        In erosion corrosion the effect of flow velocity on corrosion rate usually exhibits a breakaway phenomenon in which a maximum velocity can be withstood without removal of the protective films. Above this breakaway velocity, the corrosion rate increases at a very rapid rate. For corrosion to occur, the adhering strength of the protective film to the substrate is an important factor.
      • Impingement
        Impingement is also a type of localized corrosion that is caused by turbulence or the impingement of flow. Entrained air bubbles and suspended particles tend to accelerate the action. This produces a pattern of localized corrosion attack with directional features. The resulting pits or grooves tend to be undercut on the side away from the source of flow.
      • Cavitation
        Cavitation is mechanical damage caused by collapsing bubbles in a flowing liquid. Cavitation is caused when protective films

are removed from a metal surface by the high pressures generated by the collapse of gas or vapor in a liquid. The energy input should be sufficient to remove the protective film, and this high energy can be achieved through extreme turbulence that can cause the voids to disappear in microseconds, generating forces in the order of 700,000 kPa (100,000 psi) as the bubbles collapse.

Cavitation-related damages can be identified by deep aligned pits in areas of turbulent flow.

6. Intergranular Corrosion

As we know, the grain boundaries of metals contain impurities and are often misaligned, and intergranular corrosion (IGC) is the preferential attack at or adjacent to these grain boundaries of a metal.

The mechanism that takes place in IGC is an anodic reaction of grain boundaries, in relation to the surrounding material. The grain boundaries are anodic primarily for any of the following:
*   The impurities accumulating on the grain boundaries
*   The strain energy of the misalignment of atoms in the gain boundaries
*   The formation of precipitates caused by poor heat treatment

IGC appears as cracks. Hence, it is often called intergranular cracking and abbreviated as IGC.

Stainless steels are the material that most commonly exhibit IGC. The stainless steel IGC is metallurgically classed as general sensitization, weld decay, or knife line attack. When austenitic stainless steel is welded or hot-worked, two specific items need consideration,

A. Carbide precipitation or sensitization
B. Microfissuring, ferrite content, and Σ phase formation

The corrosion resistance of austenitic steels depends on the addition of various alloying elements of which chromium is of primary importance. If steel is heated for welding or any other fabrication or forging works in the temperature range of 800–1600 °F (425–870 °C), some of the chromium can combine with any carbon that is available to form chromium-rich precipitate. Whenever this happens, less chromium is available to resist corrosion. This type of corrosion (IGC) is more manifest in acidic environments.

The corrosion-resistant properties of austenitic steels is limited by their ability to absorb impurities without cracking during solidification and cooling from elevated temperature, as happens during welding and forging processes. Low-melting impurities are forced to the grain boundaries. Excessive amounts of these impurities in the grain bound-aries create the possibility of microscopic grain boundary flaws that are called "microfissuring." This condition is of more importance when the solidification is in a limited area, as is the case with welding and cast-ings of varying cross-sections, when the strain caused by any restraints adds to the complications.

One way to reduce the propensity for microfissuring is to alter the chemical composition of the alloy in such a way that these impurities are dispersed in a wider area, rather than accumulating in a small area. This can be achieved by introducing a greater percentage of ferrite. Fer-rite can absorb many impurities, and ferrite islands are dispersed through-out the microstructure. Increases in ferrite can decrease the corrosion resistance, but they will also definitively reduce the problem of micro-fissuring. Excessive amounts of ferrite, however, are not desired, because when metal is heated, it can transform into Σ phase. This very brittle constituent will be dispersed in all the islands of ferrite discussed above, and this dispersal makes the material weak and susceptible to brittle failure.

7. Dealloying
   The corrosion process termed dealloying involves one constituent of the alloy falling away from the solution, leaving an altered residual structure.

   The most simple example of dealloying is the separation of zinc from copper in the alloy brass. In this process one element (zinc) is anodic in reference to copper, and in favorable conditions, all zinc can fall out of the solution, leaving an altered structure of remaining copper. This can cause the failure of the structure, which can be catastrophic. Cast irons, such as gray cast iron, in buried pipeline service where SRB are present in soil stimulate this action. Cast iron has exhibited dealloying of iron and carbon to an extent that the remaining cast iron is left without any iron in it. This is also called "graphitization," as iron is the anode in relation to graphite (carbon), and it is separated out of the alloy in electrolytes such as saltwater, acidic water, and diluted acid. In such cases, one can poke a screwdriver into the thick section of the "cast iron" material that is left. Graphitization results in significant losses of mechanical strength and can be cause for failure.

8. High-Temperature Corrosion
   High-temperature corrosion is usually associated with the formation of oxide or sulfide scales. In some cases, the swelling of the wall of

the material is also noted. This is a form of material degradation that occurs at elevated temperatures. There is no electrochemical cell reaction, but instead, this form of corrosion results from a direct chemical reaction.

The basis of high-temperature corrosion is the thermodynamics and kinetics of reactions at elevated temperatures, as thermodynamics determine the tendency of metal to react in a specific environment.

CHAPTER THREE

Corrosion Control and Monitoring

INTRODUCTION

In this chapter, we briefly discuss the general issues related to the corrosion of offshore structures and pipelines. The corrosion of offshore pipelines and equipment is somewhat similar to the corrosion present in shore facilities, except that the environment is more corrosive, and monitoring is more difficult for several reasons, including proximity, reach, the application of available tools, and the variable environment, which can include the deep sea. Thus, both the external and internal threats to offshore structures and equipment are introduced here.

INTERNAL CORROSION

Internal corrosion may be due to chemical reactions on the internal surfaces of the pipelines or equipment or material loss due to microbiological reactions, which are also electrochemical. This internal corrosion of the risers or pipelines is affected by the pressure and velocities of the involved gas and liquids, among other factors. On the lay of the pipe, the inclination could also affect the internal corrosion of these pipelines. As a result, a corrosion mitigation program should include internal coating and chemical injection. Corrosion inhibitors of various types may be used. The flow phase and the lay of the pipe determine the rate of corrosion. The effect of multiphase flow at various degrees of inclination of the pipe lay also presents different challenges to the selection of the inhibitor, the concentration of its injection into the pipeline, and the monitoring system.

The slug velocity and frequency are key factors to input into the modeling of corrosion rate when determining the inhibitor type and concentration to use in a specific system. Similarly, the percentage of water in the pipe system is an important variable. In their study of CO_2 corrosion of carbon steel published in NACE publication CORROSION/93, C. De Waard et al.

proposed that no corrosion occurs if the water cut is less than 30%. Water cut is only one aspect, however, because corrosion is also dependent on factors such as superficial gas, liquid velocity, and the angle of inclination of the pipe lay.

There are several types of inhibitors, some of which are water-soluble and others are oil-soluble. The selection of a suitable corrosion inhibitor should be based on which inhibitor is most suited for the condition. The optimum concentration of the inhibitor is a matter of trial and error and requires close monitoring and adjustment of doses. Monitoring and adjusting inhibitor levels are important if one wishes to achieve an acceptable rate of corrosion for the pipeline or riser. This is a very critical operation, as failure to properly monitor and adjust can result in the rapid loss of metal to corrosion, or it can increase the cost of operation due to overdosing the system with the selected inhibitor. These challenges highlight the importance of corrosion monitoring.

Internal corrosion-monitoring techniques are suitable for topside installations, and their use is subject to periodic data retrieval for analysis, or in some cases online monitoring. Most of these techniques have very limited applications to pipelines and structures that are below the waterline or in the splash zone of a structure. Apart from measuring common points such as pipe material diameter, year of installation, and leak history, internal corrosion assessment should also collect data on the following.

- Fluid chemistry, including the presence of free water, oxygen, hydrogen, sulfur, hydrogen sulfide, chlorides, and carbon dioxide
- The reactions between the pipe material and the gases, liquids, and solids, given temperature, pressure conditions, and flow velocity
- Bacterial culture and operating stress level

MEASUREMENT AND MONITORING OF INTERNAL CORROSION

The measurement of internal corrosion and the methods used to obtain and gather these measurements have often left the operators of internal corrosion-monitoring systems frustrated because of the complicated and often difficult procedures involved. The current intrusive monitoring methods and the first generation of patch or external-type devices have also increased the cost of internal corrosion-monitoring programs. The latest developments in equipment, instrumentation, and software for the control and measurement of internal corrosion offer the operators an accurate and user-friendly alternative to what had been previously used, however. The instrumentation and software address other parameters that cause the onset of or increase in corrosion, so that the operators can relate corrosion to the events that cause corrosion.

BASIC INTERNAL CORROSION MONITORING TECHNIQUES

1. Coupons (C)
2. Electrical resistance
3. Linear polarization resistance
4. Galvanic
5. Hydrogen probes and patches (H_2)

In addition to the above-listed techniques, some new corrosion-monitoring methods are being tested in the laboratory and in the field. These methods include:

1. Surface activation and gamma radiometry
2. Impedance (EIS)
3. Electrochemical noise
4. Acoustic emission
5. Real-time radiography
6. Real-time ultrasonic
7. Hydrogen patches (H_2)

INTERNAL CORROSION MONITORING DURING PIGGING OPERATIONS

The above-listed monitoring techniques are grouped into either intrusive or nonintrusive methods. Intrusive methods usually involve inserting a coupon or probe into the pipeline or piping system at the pump or compression stations, gas and oil separation platforms, and other topside equipment or related piping arrangements. In order to accomplish this insertion, the operation must occur either at full operating pressure, or operators must shutdown the process line for the insertion and removal of the monitoring sensor. Another alternative is to place the sensors on a bypass with a block-and-bleed capability, but many feel that the environment in which the corrosion occurs differs from the bypass, and hence, the data received from bypass does not reflect the actual conditions of the equipment and piping systems. For this reason, most operators do not use bypasses unless there is no alternative.

With flush-mounted probes, there is no interference with the operation of the equipment or the passage of the pig, if the topside piping system is designed for pigging. A standard access fitting can be installed at any point around the circumference of the pipe that allows the probe or coupon to be properly placed for monitoring corrosion. By obtaining proper calculations of the pipe-wall thickness, the weld gap, and the design of the sensor, sensor holder, and plug within the access fitting, operators can install sensors that are truly flush with the inside of the pipe. In addition to allowing the passage of pigs, this design places the sensor at the pipe wall where corrosion is occurring. The potential for corrosion near the middle of the gas or liquid stream has little or no bearing on the general corrosion occurring on the pipe wall itself.

Nonintrusive methods do not interfere in any way with operations. Unfortunately, at this point in time, only a few nonintrusive methods are field-proven, and equally few give the operator adequate or complete internal corrosion monitoring data. For instance, while hydrogen patch devices work well in many environments, there are many applications where they do not seem to work at all.

EXTERNAL CORROSION

For the sake of this discussion, offshore structures can be divided into structures that are above the water level and those that are below water. The above-water structures and equipment are often protected from external corrosion by suitable coating, selection of CRA, or a combination of both methods. The below-water group is divided into structures that are in the splash zone and structures that are fully submerged in water. These two areas are generally coated with a high-efficiency coating system. In addition, submerged structures are protected from corrosion by making them cathodes in relation to a set of sacrificial anodes placed in the sea water, which is the electrolyte in the corrosion cell. We will discuss these methods in more detail in the subsequent chapters.

Anode Materials and Testing

CHAPTER FOUR

Anode Materials

INTRODUCTION

Anodes are made out of alloys that are more electrochemically active than the material they are intended to protect from corrosion. As a result of interactions with surrounding structures and fluids, the anode depletes during the lifetime of the designed structure. The anodes that are used to prevent corrosion in cathodic design systems are called sacrificial anodes.

The principle of the sacrificial anode is the same as the principle of cathodic protection, wherein metal surfaces come into contact with electrolytes, leading to an electrochemical reaction called corrosion. Metal in seawater is an example of this principle, as exemplified by a steel structure coming into contact with electrolytes in the ocean. Under normal circumstances, the steel will react with the electrolytes and begin to corrode, growing structurally weaker and disintegrating over time. However, if the electrolyte contains another metal that is more active than the steel, then the more active metal will corrode preferentially, thus protecting the steel from corrosion.

As described above, these more active metals are the sacrificial anodes positioned in the same electrolyte in order to corrode preferentially compared to the structure of the interest. The sacrificial anodes that are used in offshore cathodic protection are intended to be consumed over the lifetime of the protected structure.

For example, in cathodic protection designs where anodes are buried to protect land structures, a special backfill material surrounds the anode in order to insure that the anode produces the desired output. The backfill material acts as regulator of uniform, constant, well-distributed current output.

Finally, the sacrificial anode works by introducing another metal surface that is a more electronegative and anodic surface. The system allows the current to flow from the newly introduced metal, as an anode, and the protected

47

metal then becomes cathodic, creating a galvanic cell. The oxidation reactions are transferred from the metal surface to the galvanic anode and are thus sacrificed in favor of the protected metal structure.

ANODE MATERIALS

In relation to standard carbon steel, which is used for the construction of structures and piping components placed in the seawater, materials such as aluminum, zinc, magnesium, and their purpose-designed alloys are more active, and hence, they corrode preferentially to the steel structure, protecting it.

The current capacity and operating voltage of the sacrificial material are the two most important properties to be considered in selection of anode material. The current capacity of an anode is its ampere hour per unit output. The anodes are tested to give consistent output in densities from as low as 860 mA/M^2 to 15,100 mA/m^2. If an anode is well cast, the current capacity values should not be significantly affected by changes in environment and operational variables.

The operating potential of an anode material is another factor that gives the driving voltage for the current to flow and protect the structure. This property must balance the high potential to drive the current, but it must not be so high that it depletes the life of the anode (Figure 4.1).

Metals with the described properties are used as anode materials, but they are often alloyed to improve their performance in terms of current output. For example, the standard reduction potential of zinc is about -0.76 V. The standard reduction potential of iron is about -0.44 V. This potential difference between two metals is the key to cathodic protection using sacrificial anodes, and it creates the driving voltage of a zinc anode if used in this form. The driving voltage should not adversely affect the output current densities, should be constant throughout the life of the installed anode, and it should not polarize.

The consistency of both the current capacity output and the potential behavior are key to a good anode material. The above discussion about using pure metals and alloys as anodes suggests that anodes must be tested to ensure that they meet the desired current output.

Quality control in the manufacturing of anodes depends on the specific requirements of the project that will employ those anodes. The testing and quality control of anodes is detailed in Chapter 5. However, some common aspects of this process are:

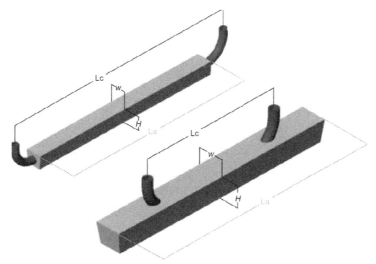

Figure 4.1 Typical anode type shows the steel core and various dimensions of an anode.

- Chemical composition
- The required electrochemical output (ε) measured in amp-h/kg
- The closed circuit anode potential of the anode (measured in volts)
- The anode utilization factor

Alloying is intended to improve the performance of the anode, and anode performance is determined by the values obtained for the four aspects listed above.

The manufactures have very secretly guarded the compositions of their products, and they have trade names for specific products. These specialty alloys are cast by fine-tuning the elements included in the alloy, as well as the amount of each element in the composition. However, most of the anode castings used to protect offshore structures are composed of aluminum- and zinc-based alloys. The typical chemical composition for aluminum- and zinc-based alloy castings are given in Table 4.1.

ANODE MANUFACTURING

The manufacturing process for anodes is a structured activity in which the following are detailed in the procedure.

- The quality and source of the raw material
- The melting and casting practices, including temperature and mold used

Table 4.1 Typical Chemical Composition Analysis of Al and Zn Alloy Anodes

Type of Anode ↓	Zn	Al	In	Cd	Si	Fe	Cu	Pb
Al based alloy	2.5–5.75	Balance	0.015–0.040	≤0.002	≤0.12	≤0.09	≤0.003	none
Zn based Alloy	Balance	0.10–0.50	none	≤0.07	none	≤0.005	≤0.005	≤0.006

- The quality and specification of the anode core material
- The alloy's constituent elements and their percentage composition
- The size and shape of the anode detailed in a drawing that includes the weight of anode, the inserts used, and other relevant information
- The inspection and testing of the anode produced

The anode production process begins with the production of sample anodes, using the agreed upon parameters discussed above, followed by the necessary tests. These tests include destructive and nondestructive tests to determine the properties of the anode. The successful results of these tests determine the validity of the casting process for the anode.

STEEL INSERTS

The inserts in the anode are the steel bodies cast into the anode material. These inserts are often made out of a ductile grade of steel that can be welded to the structure to attach the anode.

CONNECTIONS TO THE ANODE

A low resistance, mechanically adequate attachment is required for good protection and resistance to mechanical damage. In the process of providing electrons for the cathodic protection of a less active metal, the more active metal corrodes. The more active metal (anode) is sacrificed to protect the less active metal (cathode). The amount of corrosion depends on the metal being used as an anode, but it is directly proportional to the amount of current supplied.

Sacrificial anodes are normally supplied with lead wires or cast-in straps to facilitate their connection to the structure being protected. The lead wires may be attached to the structure by welding or mechanical connections (Figure 4.2). These wires should have a low resistance and should be

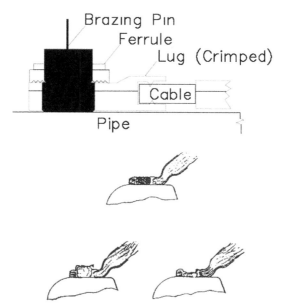

Figure 4.2 The picture of a lug welding.

insulated to prevent increased resistance or damage due to corrosion. When anodes with cast-in straps are used, the straps can either be welded directly to the structure or used as locations for attachment. A cross-section of a welded connection on a structure is shown in Figure 4.2. The resistance between the connection and the structure is less than $0.1\ \Omega$.

CHAPTER FIVE

Anode Testing and Quality Control

INTRODUCTION

Testing anode castings for quality and output is an essential part of the bulk anode manufacturing process. The testing should be carried out as part of the manufacturing process control regime. The testing should also be conducted in accordance with the industry standards established by NACE in the testing method document NACE TM 0190.

SAMPLING AND TESTING

The producer and purchaser should agree on the sampling, testing, and acceptance criteria prior to manufacture. The NACE has developed a document that outlines recommended practices (NACE RP 0387), including detailed methods for sampling and testing anodes. These recommended practices may be used to help the producer and purchaser develop a specific sampling method. The sampling of the product for testing is an important part of the testing protocol. A test specimen from each batch of the cast should be tested. The selected specimen should be identified with the batch and cast number it represents. The selected specimen of specified dimension and weight is then immersed in the test solution of sea water or artificial sea water. The ASTM D 1141 has detailed directions on the preparation of seawater solution for testing. The weight is recorded to 0.1 mg accuracy, and the solution temperature is maintained between 17 and 23 °C during the testing.

The test specimen is suspended in an uncoated steel pipe, and this pipe serves as the cathode in the test setup. The surface area of the cathode steel pipe in the electrolyte should be about 20 times that of the anode surface. A DC positive-end terminal is connected to the anode, and the negative terminal is connected to the steel pipe cathode in the electrolyte to initiate a galvanostatic reaction. The current discharge is accurately measured and recorded to the accuracy of ±2%, using an electronic device or a copper coulometer current integrator. During the testing period, the current is varied to different current densities, and the densities maintained within a precise tolerance of ± 0.1 mA/cm^2. Each variation of the current density is held for at least 24 h, and the total test is carried out for four different density levels, starting with the current density of 1.5 mA/cm^2. The initial current density is then reduced to 0.4 A/cm^2, raised to 4.0 mA/cm^2, and reduced to 1.5 mA/cm^2. The total time for the test is about 96 h.

THE TEST ANALYSIS AND THE TEST RESULT

After the final stage of the test is completed, the specimen is removed from the bath and cleaned of the corrosion product. There are different solutions and methods for cleaning an aluminum alloy or zinc alloy specimen. Aluminum alloy specimens are cleaned to remove corrosion products, using a 20-g chromium trioxide and 30-ml concentrated phosphoric acid solution at 80 °C temperature for about 10 min. Zinc alloy specimens are cleaned by immersing them for about 2 h in a saturated ammonium chloride solution at ambient temperature. The cleaned and dried specimens are then weighed to 0.1 mg accuracy, and the recorded change in weight (Δm) is compared with the initial weight recorded at the start of the test. This delta weight is used to compute the electrochemical capacity of the anode. The electrochemical capacity (ε) of the anode material is calculated using following relationship between the electric charge (Q), measured in ampere hour, and the change in the weight (Δm) of the specimen in kilograms.

$$\varepsilon = Q/\Delta m$$

Tests on a number of specimens are recorded and evaluated to meet the specified criteria. It may be noted here that, at higher current output, the hydrogen output is also higher, and this fact must be kept in mind when selecting an anode in conjunction with high-strength steels and certain alloys.

Cathodic Protection of Offshore Structures

CHAPTER SIX

Offshore Structures

INTRODUCTION

The use of cathodic protection can be traced back to 1824, and its application preceded Sir Humphrey Davy's theory explaining the involved phenomena. Davy successfully used iron, zinc, and tin to protect copper hull cladding on naval vessels exposed to seawater. This was the first application of cathodic protection (CP). By 1945, CP was commonly applied in the United States, especially by the rapidly expanding oil and natural gas industry.

CP is now well established, and it provides effective corrosion control for a large variety of immersed and buried metallic structures, as well as reinforced concrete. In simple terms, CP works by preventing the anodic reaction of metal dissolution from occurring on the protected structure.

For the purposes of our discussion, offshore structures consists of pipelines, fixed and floating platforms, tension leg platforms (TLPs), SPARs, floating production storage and offloading facilities (FPSOs), pipeline end manifolds, pipeline end terminals, and various other supporting and essential constructions. All these are metallic constructions of various types, and they are either fully or partially submerged in a common electrolyte; that is, in most cases, seawater. Some of these structures or their parts are buried in the mud on the sea floor, while others are in the seawater or the splash zone, where they are sometimes under water and sometimes exposed above the water level.

Often these structures are built on steel legs anchored directly onto the seabed, with a set of decks that support space for oil and gas separation, drilling rigs, production facilities, and crew quarters. Such platforms are, by virtue of their immobility, designed for very long-term use. Various types of structure are designed for different service and use, and these design types include steel jacket, concrete caisson, and floating steel. Steel jackets are vertical sections made of tubular steel members that are usually piled into the

Figure 6.1 Typical SPAR.

seabed. In floating structures, these members are often used as a flotation chambers to stabilize the structure.

Figure 6.1 shows a typical SPAR.

As with TLPs, SPARs are moored to the seabed, but, while TLPs have vertical tension tethers, SPARs have more conventional mooring lines. A SPAR has more inherent stability than a TLP because it has a large counterweight at the bottom and does not depend on the mooring to hold it upright. It also has mobility and the ability to adjust the mooring line tensions using chain-jacks attached to the mooring lines. The adjustability of the moorings allows horizontal movement, so that a SPAR might be positioned over wells at some distance from the main location of the platform.

SPARs are primarily designed as:

- A conventional one-piece cylindrical hull
- A truss SPAR in which the midsection is composed of truss elements connecting the upper buoyant hull (called a hard tank) with the bottom soft tank containing permanent ballast
- A cell SPAR, which is built from multiple vertical cylinders.

The typical advantage of a SPAR is that it has relatively reduced hull weight, high damped heave response, reduced loop current load on the hull, relatively shallow draft, and reduced cost of fabrication compared to other structures.

The mooring system may consist of legs with permanent anchor piles, anchor chains, jacketed steel spiral strand wires, or platform chains.

The hull (hard tank) is the cylindrical steel structure that gives the buoyancy to the SPAR. Often the hard tank or the portion of it that is at water level is a double-walled construction. To suppress and reduce the vortex-induced vibration, the hard tank is fitted with steel strakes.

The SPAR trusses are X-braced in a manner similar to X-bracing on offshore jackets, and the trusses serve as a rigid structure between the hard tank above and keel tank below. The trusses are often fitted with heave plates to dampen the heave response.

Finally, at the bottom of the structure, the keel provides buoyancy, and it is flooded to upend the hull.

The external surfaces of the structure are exposed to seawater, and as a result, these surfaces require protection from corrosion, at least in part, by a cathodic protection design. To ensure adequate protection, the surface areas of all exposed structural members are calculated during the design of the CP system.

Despite their similarities, all other types of offshore structures have designs that are specifically tailored to their intended purposes, and the nature and use of the structures determine which exposed surfaces may be coated or painted and which need to be safeguarded via cathodic protection.

Figure 6.2 graphically shows various types of offshore structures and their typical locations in relation to the shore and depth of waters in which they are installed.

Figure 6.3 shows a typical semisubmersible platform.

Figure 6.4 shows the general layout of various gathering, flow lines, and pipeend equipment on the sea floor.

Figure 6.2 Various types of offshore structures and their typical locations.

Figure 6.3 Shows a typical semisubmersible platform.

Figure 6.4 General layout of various gathering, flow lines, and pipe end equipment on the sea floor.

As is seen from above pictures and brief description, offshore structures involve massive investments, and they also pose potential risks for both people and the environment, if they fail. One of the potential reasons for the structures' failure can be corrosion, highlighting the importance of corrosion protection and its design.

For offshore structures, the main form of corrosion is external corrosion. This corrosion distribution results from the external environment, and the reactions between the structural materials and the environment. The conventional approach to the prevention of external corrosion is to cover the exposed surfaces with a high-efficiency coating, while making the structure a cathode in relation to an anode placed in the electrolyte formed by the seawater. The creation of a cathode and an anode implies that an electric circuit is created so that a net current flows to the structure in order to make it a cathode. In an offshore environment, this net electrical flow is primarily achieved through the use of sacrificial anodes. The limited use of impressed current systems is also practiced, and this method will be discussed in subsequent chapters. Several international organizations have developed specifications that govern the design of cathodic protection, coating, and other forms of corrosion prevention, and some of these specifications are listed in Section 5 of the book for reference.

These specifications provide excellent guidelines for designers hoping to meet basic essential requirements when creating a cathodic protection system. As a result, these specifications should be referenced for project-specific requirements.

The general objective of a cathodic protection system is to polarize the steel structure to an electrical potential generally between -800 mV and -1100 mV, as measured against a Silver/Silver chloride/Seawater (Ag/AgCl/Seawater) reference electrode. This polarization is maintained throughout the designed life of the structure. However, some other polarization criteria for steel have been developed for more specific environments.

a. More negative than -800 mV/Ag/AgCl/seawater in aerated seawater

b. More negative than -900 mV/Ag/AgCl/seawater in a seawater environment with high numbers of sulfate-reducing bacteria

c. More positive than -1100 mV/Ag/AgCl/seawater in seawater if the yield strength of the structural steel does not exceed 550 MPa (about 80 ksi)

The polarization potential for other materials is different from the steel discussed above. EN 12473 details some of these polarization voltages. For example, some high-strength alloys and corrosion-resistant alloys (CRAs) tend to have a high risk of hydrogen-induced cracking, and as such, they cannot be overpolarized. So, a different criterion of protection is applied.

The bare (noncoated) material should not be polarized to a more negative potential than -1.05 mV in reference to the Ag/AgCl/seawater reference electrode.

The yield strength of steel is a very important factor in determining how much protection current can be imposed during the cathodic reaction, while preventing the production of atomic hydrogen. As with most CRAs, the control of hydrogen production is essential for steels with higher yield strengths and hardness levels approaching 300 HV_5 (307 HB or 32 HRC).

The environment of each of field is unique in terms of current density demand. This is especially true in case of seawaters, which often vary in salinity and temperature. Table 6.1 provides the generally acceptable current densities for cathodic protection systems in various seas around the world. There are various international organizations that issue specifications that have addressed these variables, and they all present somewhat differing numbers. So, it may be appropriate to note that none of the specifications are absolutely right or wrong, because the estimated current density demands totally depend on the data the organizations recorded, which in turn depend on when and where they collected that data. In more involved

Table 6.1 Typical Current Densities

Geographical Location (Area)	Current Densities (m/Am^2)		
	Initial or Polarization	Maintenance	Final or Repolarization
North Sea: North of 57 °N (0–30 m depth)	200	100	130
• 30–100 m	170	80	110
• 100–300 m	190	90	140
• 300 m	220	110	170
North Sea: South of 57 °N (0–30 m depth)	250	120	170
• 30–100 m	200	100	130
• 100–300 m	220	110	170
• 300 m	220	110	170
Mediterranean Sea	150	70	90
Gulf of Guinea Shallow Waters	130	90	100
Deep Waters	250	100	100
Gulf of Mexico	110	55	75
US West Coast, Cook inlet	150	90	100
Persian Gulf	130	70	90
Australia	130	90	90
South China Sea	100	35	35
Indonesia	110	60	75
West Africa	130	65	90
Buried Zone (ambient temperature)	25	20	20
Splash Zone	Typically a 10% increase over the above recommended current density is taken		

design applications, designers should consider measuring the actual current densities in the area where the structure will be placed.

Other key variables in cathodic protection design are:

1. The design life
2. Surface conditions: Presence of coating, type of coating applied, coating breakdown factors
3. Use of mud mats, skirts, and piles
4. Current drain to wells and anchor chains
5. Electrolyte resistivity
6. Type and quality of sacrificial anode

A typical cathodic protection design for a deep-sea offshore structure includes a number of different structures, as described in paragraphs above.

CATHODIC PROTECTION OF FIXED OFFSHORE STRUCTURES

The cathodic protection of an offshore structure starts with the concept design of the structure. An essential part of the design process is devoted to corrosion control, and during this phase, everything is considered, including the selection of material, calculations of thickness, the environment where the structure will be installed, provision of corrosion protection secondary to coating, and the method of installing CP anodes.

In the chapters that follow, we address these aspects of corrosion protection, with emphasis on the cathodic protection of offshore structures.

Structural Design for Corrosion Control

The design of a corrosion control system starts with the design of structure itself. This approach reduces the possibility of including design features that may become detrimental to a structure's life and performance. By their very nature, some design features may be prone to promote corrosion or may lead to corrosion failure. For example, certain weld-designs that could be the cause of corrosion are listed below.

• Stitch welding
• Weld-designs that include backing strips
• Weld ends that aren't rounded to create a seal
• Rough surface welds that are not dressed to remove the possibility of stress points

- A possible crevice in a weld that could become a node for corrosion and possible failure

The design of the offshore structure itself could present unique issues that increase the likelihood of corrosion. Potentially corrosion-prone design features include T-K-Y-type joints, joint location, member materials, joint fitups that overstress adjacent materials, weld designs that lead to over-welding, welds with improper high-low positioning and poor transitions, pipe-ends with or without a dead leg, and the absence of seals for the ends of members to be flooded during the installation process. The very location of a structural member could present corrosion challenges.

The structure is installed in the sea water. So, some of the structure is submerged, and some part of it is above the water level, exposed to atmospheric elements; however, there will be some parts of the structure that are in a zone where they are intermittently in and out of seawater, primarily due to shifting water levels resulting from tidal movement. These three areas of varying exposure are called zones, and these zones are an important part of the cathodic protection design, as they each require different approaches. We briefly define what these three zones are.

ATMOSPHERIC ZONE

The atmospheric zone is the section of the structure that is above the splash zone, and by its very definition, this zone is not wetted or affected by the rise of tidal waves; hence, it is not in the electrolyte. The corrosion control of this zone is achieved through the application of a suitable coating system.

Given that this zone is not in contact with the electrolyte (seawater), the coating does not need to be complimented by cathodic protection, the structure design essentially tries to minimize the exposed steel surface area in this zone. Minimizing the exposed surface area is primarily done through the use of tubular members. Clean welds with good profiles that blend into the parent metal then prevent any crevices that might promote corrosion, and the boxing in of the steel structures can accomplish the same goal. Substituting steel with corrosion resistant materials and nonmetals wherever possible is another example of designing a structure with a reduced tendency to corrode. Planners should avoid designs in which dissimilar metals are joined, however, or they should provide insulation between dissimilar metal joints.

This approach reduces the possibility of detrimental design features that by their very nature are prone to corrosion.

SPLASH ZONE

The splash zone is the section of the structure that is intermittently in or out of seawater during the structure's service life. Tides and wind are often the responsible for wetting this section. The range of the zone varies with the height of the sea's rise and fall due to daily tide cycles in the specific geographic area. For example, the range of the splash zone in the Gulf of México is about 6 ft., but it is about 33 ft. in North Sea and about 1 m (3.32 ft.) in the North China Sea. The corrosion protection of material in this zone requires a different approach than the conventional approach discussed for the submerged zone.

SUBMERGED ZONE

The submerged zone is the section of the structure that is below the lowest end of the splash zone and is always below the sea level. From the cathodic protection point of view, this portion of the structure is always in the electrolyte. The corrosion protection of this zone is achieved through a well-designed program that includes the application of a high-performance coating system supplemented by an equally well designed CP system.

To reduce corrosion risk, the structural design preferably uses tubular construction, as tight or recessed corners and crevices are difficult to protect in this zone. A stress-reduction design is also encouraged to minimize fatigue stress, and where required, the design should also relieve the stress placed on welded members.

Structure-pilings, skirt-piles, and centralizers are considered to be part of the structure for CP system design. When these members are encased within the jacket legs, they are bonded to the legs by welding, and their surface area is to be considered while calculating the required polarization current. The parts that are not welded must be electrically bonded to the structure to ensure that they are part of the CP system. After electrical bonding, non-welded members can be considered one with the rest of the structure during calculation of polarization current and the determination of anode mass and quantity.

Although still in the submerged zone, members below the mudline generally experience a very low corrosion rate, as there is scant oxygen or turbulence to act on the steel; however, members buried in the seafloor should be made part of the total CP system.

Chains are parts of the structure that straddle the splash and submerged zones, and they are often designed with twice the corrosion allowance than the corrosion rate. The splash zone requires a higher corrosion allowance, and there is no CP provided to chains. However, where possible, designers should attempt to keep the lay of the chain within the effective radius range of the CP current, thus providing the chains with some protection.

Unlike chains, galvanized steel wire ropes are protected by an outer sheathing of polyurethane filled with grease to prevent the ingress of water. This sheathing provides the galvanized wires with sufficient protection from corrosion. However, a good wire rope design will include a sacrificial zinc wire within the rope construction as secondary protection.

CRITERIA FOR CP

Criteria refer to measured data indicating if adequate cathodic protection is provided to a metallic structure. These criteria can be determined by visual inspection, by measurement of pipe wall thickness, or by use of internal inspection tools. Given that these methods lack universal application or quantitative evaluation, additional measures are devised and used for verification of cathodic protection.

For example, over the years, several laboratory experiments have been carried out to establish measurement methods, and the resulting acceptance criteria are fixed or derivable values that can be universally used. Singularly or in combination, these criteria and any one of the methods discussed above can be used to determine the adequateness of the protection provided.

External corrosion control can be achieved through various levels of cathodic polarization. These variations are a function of the environment in which the structure is located. The accepted criteria for cathodic protection are:

- A negative polarized potential of -850 mV (-0.85 V), with reference to a $Cu/CuSO_4$ solution electrode.
- A negative cathodic potential of at least -850 mV (-0.85 V) when the cathodic protection current is on (This criterion presupposes that sound engineering calculations are made to evaluate the drop in voltage across the structure and electrolyte. The physical environment is evaluated, and cathodic protection data is regularly monitored.)

- A shift in cathodic polarization of minimum 100 mV between the structure and stable reference electrode connecting the electrolyte

These established criteria may not be adequate in certain cases. A different set of criteria should be determined if any one of the following conditions exist. Sound engineering evaluation should be made to determine and establish the following protective criteria (Table 6.2).

- The environment (electrolyte) where the structure is electrically in contact with bacterial activities
- The environment (electrolyte) where the structure is in contact with environmental sulfides
- The environment (electrolyte) where the structure is in an acidic environment
- The environment (electrolyte) where the structure is in an environment with an elevated temperature
- A structure to be protected containing dissimilar metals

Table 6.2 Rrecommended Sea Water Potential for Different Metals in 30 Ω-cm Seawater, Where Salinity is Between 3.2% and 3.8%

Material to be Protected	Minimum Negative Potential (V)	Maximum Negative Potential (V)	Remarks
Carbon steel in aerobic environment	−0.80	−1.10	Where the steel $\sigma_{smys} > 80$ ksi, the most negative potential should be evaluated to avoid hydrogen embrittlement
Carbon steel in anaerobic environment	−0.90	−1.10	
Austenitic steels where PREN is ≥40	−0.30	−1.10	
Austenitic steels where PREN is <40	−0.50	−1.10	
Martensitic steels	−0.50	See remarks	Depends on whether the strength
Duplex stainless steel and nickel alloys	−0.50	See remarks	of material, residual stress, and stress in service make the material susceptible to hydrogen embrittlement, but the potential may be limited to no greater than −0.80 V

The polarized potential that would generate excessive hydrogen should be avoided on all metals. Molecular hydrogen ingress and hydrogen–induced failures are possible on high-strength and high-residual-stressed metals. Similar conditions apply to martensitic steels and Duplex steel and nickel alloys.

DESIGNING A CP SYSTEM

The process of designing a cathodic protection system involves calculations of current demand for three distinct stages of the structure's life. Various factors influence the determination of current demand. Temperatures, turbulence, resistivity, chlorine level, and depth are some of the main factors that affect current demand for polarization and its maintenance throughout the life of the structure. These important factors are discussed through explanations, data tables, and graphs.

TOTAL PROTECTIVE CURRENT REQUIRED

The cathodic protection of any structure is often designed to protect the structure over its designed life. The intended duration of the protection implies that the design should ensure that the polarization current is effectively available throughout structures' design life for continued maintenance of cathodic polarity of the structure. The demand for polarization current is not same through the structure's designed life, and the current output of installed anodes also varies during this period. As a result, design calculations for various stages of design life are made separately to estimate an acceptable level of polarization current. There are three distinct stages in the life of a structure as it relates to cathodic protection design and current demand, and these are discussed further.

INITIAL OR POLARIZATION CURRENT

The CP system is required to have an initial current density to polarize the new steel to and make it a cathode as fast as is possible after the structure is immersed in the seawater. This is called initial current density, and often this first stage has the highest level of current demand. The process involves using Ohm's law to determine the resistance, while calculating the required current demand using Dwight's, Crennell's, or McCoy's equations. These calculations are based on the original dimensions of the sacrificial anodes. The sample calculation below uses these equations.

MEAN OR MAINTENANCE CURRENT

The second step in the design of the CP system is to calculate the maintenance current density, which is also referred to as the mean current density. Mean current is used to determine the total weight of the anode that will be needed to maintain the current required to protect the structure over its design life.

FINAL CURRENT

Finally, the third calculation is carried out to establish the final current density. This current density is intended to ensure that enough current remains to provide protection at the end of the structures' life and also when the anode is depleted and nearing end of its own life. This calculation is similar to the initial current calculation, with the depleted dimensions of the anode used in place of the original dimensions.

FACTORS TO BE CONSIDERED FOR CURRENT CALCULATIONS

The principal factor for the effective design of a CP system is the selection of the correct design current density. The polarization current for the environment should be calculated and provided for. The depolarizing agents in the environment, which is often the open sea, should be considered, and compensation should be made to maintain the required polarization current density. The major depolarizing agent found in the open sea is dissolved oxygen. The temperature of the water also plays significant role in the presence of oxygen, with lower water temperature allowing more oxygen to be dissolved. Water turbulence and lateral flow increase the amount of dissolved oxygen in water as well, and this oxygen is transported to the structure surface.

In colder environments, the damage to the coating and substrate increases with the presence of glacial silt and scouring, increasing the depolarization activity. Lower temperature also increases the resistivity of water as can be seen from Table 6.3.

The reduction in polarization current demand is produced by calcareous deposits, as these deposits reduce the amount of oxygen reaching to the substrate. The shift of the pH level into the alkaline range creates the calcareous

Table 6.3 Sea Water Temperature and Effect on Resistivity

Temperature:→ °F (°C)	32(0)	40(4.4)	50(10)	70(21)	75(24)
Resistivity (Ω-cm)	35.1	30.4	26.7	21.0	19.4

deposits in seawater. The positive effect of calcareous deposit in reducing the demand for polarization current is also adversely affected by the change in water temperature.

Higher initial current densities are required for the development of calcareous deposits. Once the calcareous deposits are formed, the current demand rapidly drops. Field experience and laboratory tests have shown that calcareous deposits develop faster when a steel structure immersed in seawater is rapidly polarized to the potentials of −9 to −1.0 V, measured with reference to an Ag/AgCl electrode reference electrode in sea water. It is also a practical observation that, when a new steel structure is launched in the sea, the initial measured potential difference between the sacrificial anode and the unpolarized steel structure is about 0.45 V. After the calcareous deposit is formed, this potential difference drops down to ≥0.25 V, and this corresponds with the lower current density demand. The Design criteria for the cathodic protection of offshore structures are given Table 6.1, and the point to note is that these variations are also supported by the change in the seawater temperature and turbulences in specific environments. The following table gives a list of recorded temperature, turbulence, and water resistivity of various seas around the world. The resistivity of water is also affected by the chlorine level, and the resistivity indicated in Table 6.4 is based on 20 ppt chlorinity in the water. Note that the values in the table may change with different regions of the same sea, and given the differences in local environments, an actual survey of temperature, chlorinity, and resistivity should be conducted for accurate design calculations.

As we have noted above, the current density is strongly related to the temperature of the water, and it also varies with the depth of the water. The water-depth average temperature for each interval should be used for design calculations. If the structure straddles different temperature zones where the temperature difference is over 5 °C, then suitable adjustments should made to create a new zone, and new calculations should be conducted to establish the current density demand for those different new zones. Similarly, the resistivity of water also differs with the depth of the water. As was done for temperature, designers should attempt to position the structure to minimize resistivity differences in the zones to which it is exposed, while

Table 6.4 Various Sea Water Resistivity and Turbulence

Sea	Water Temperature (°F, °C)	Water Resistivity (Ω in., Ω cm)	Typical Turbulence	Typical (Avg.) Lateral Water Flow
US West coast	59 (15)	9.45 (24)	Moderate	Moderate
Gulf of Mexico	72 (22)	7.87 (20)	Moderate	Moderate
Brazil	59–68 (15–20)	7.87 (20)	Moderate	High
North Sea	32–50 (0–10)	10.24 to 12.60 (26 to 32)	High	Moderate
Persian Gulf	86 (30)	5.90 (15)	Moderate	Low
South China Sea	86 (30)	7.09 (18)	Low	Low
West Africa	41–68 (5–20)	7.97–11.81 (20–30)	Low	Low
Australia	54–64 (12–18)	9.10–11.81 (23–30)	High	Moderate

repeating calculations with new values to more accurately define the new resistivity.

Process of polarization should produce a linear graph of steel potential and current density, in which the slope of the line is equal to the overall circuit resistance of the CP system. The design slope and the maintenance current density are related, and their specific relationship can be established by on-site measurements and review of data collected from the specific location.

DESIGN CALCULATIONS

The factors involved in the design of a CP system for offshore structures relate to conditions that will support the structure's protection. The factors to be calculated include:

- Surface area
- Current density
- Current drain
- Anode mass
- Number of anodes
- Anode resistance
- Distribution of anodes
- Electrical continuity

For any offshore structure, especially platforms and substructures, a conceptual design is prepared to identify the type of anode and method of attachment, thus addressing the demands of installation operation activities. The conceptual design then feeds into the detailed design.

The detailed CP design requires further calculation, for which following input information is essential,

* Design life for the structure and cathodic protection
* Project design basis information on depth, water temperature, seawater salinity, mud-line, anode resistance, and related factors
* Fabrication details and drawing of the structure and substructures
* Electrically connected interfaces of components and systems
* CP protected structures, such as pipelines, in proximity to the structure
* Survey reports
* Type of coating applied and coating breakdown factors (refer Tables 6.5 and 6.6)
* Information on available types and sizes of anode
* Anode utilization factor

SURFACE AREA CALCULATION

For the designing the CP system, it is essential to know the surface area to be protected. The large and complex geometry of the structure is divided into different CP units. The surface area for each CP unit's geometrical subsections are determined and combined. Normally, for the purposes of the CP system, the structure is also divided according to depth zones, temperature zones, and physical restrictions or electrical isolation.

Table 6.5 Recommended Coating Breakdown Factors

Coating Type	Applied Concrete Coating	Maximum Continuous Operating Temperature (°C, °F)	Constant-a	Constant-b
Glass fiber reinforced asphalt enamel	Yes	70 (160)	0.0003	0.0001
Glass fiber reinforced coal tar enamel	Yes	80 (176)	0.0003	0.0001
Single or dual layer FBE	Yes	90 (194)	0.01	0.0003
3-Layer FBE or PE	Yes	80 (176)	0.0001	0.00003
3-Layer FBE or PE	No	110 (230)	0.0001	0.00003
Multilayer FBE or PE	No	140 (284)	0.0003	0.00001
Polychloroprene	No	90(194)	0.0001	0.0001

Table 6.6 FJC breakdown Factors

Type of FJC	Maximum Temperature (°C, °F)	Constant-a	Constant-b
None		0.3	0.03
Adhesive tape or heat shrink sleeve with PVC or PE backing or with mastic adhesive.	70(160)	0.1	0.01
Heat shrink sleeve backing plus adhesive in PE, LE primer (some in-fill)	70(160)	0.03	0.003
Heat shrink sleeve backing plus adhesive in PE, LE primer (no in-fill)	110(230)	0.03	0.003
FBE with infill	90(194)	0.03	0.001
FBE with PE heat shrink sleeve with infill	80(176)	0.01	0.0003
FBE with PP heat shrink sleeve No infill	140(284)	0.01	0.0003
FBE, PP adhesive and PP wrapped, extruded or flame sprayed, no Infill	140(284)	0.01	0.0003
Polychloroprene	90(194)	0.01	0.0003

Once the structure drawing is marked with the planned subdivisions, the surface area is calculated for that CP unit. The parts that are coated are calculated separately from the parts that are not coated, or they are identified separately with notations that include the type of coating applied and the surface temperature if different from the normal design value. In the process of area calculation, it is possible that complex shapes will present serious challenges to the accuracy, and for cathodic protection design, a tolerance of -5% to $+10\%$ is normally acceptable in area calculations.

It is practical to use the drawing reference for each area calculation, as this allows for the reverification and tracking of each subsection of a CP unit.

The effective design of a CP system also requires designers to know the *current density* (i_c) required to achieve the desired level of protection. The current density is the cathodic protection current required per unit area, and it is measured in amperes per square meter (A/m^2). Densities are calculated for the initial (i_{ci}) and final (i_{cf}) current demands. These values are used to determine the anticipated current density demands, by determining the number and sizes of anodes to use on bare steel surfaces. The effect of the coating system is accounted for by using the coating breakdown factors given in Tables 6.5 and 6.6.

INITIAL CURRENT DENSITY

The initial current density (i_{ci}) is the current demand per unit area that will provide the effective polarization of bare metal that may have some scale rust or mill scales. As stated above, the effect of any coating is taken into account by using the coating breakdown factor. In effect, the initial current density is higher than the density required for maintenance or the final stages of the structures' life.

MAINTENANCE (MEAN) AND FINAL CURRENT DENSITIES

As the initial current is applied to the structure, the polarization of the structure takes about 4–8 weeks. After that, the current demand gradually drops to a level that is far lower than the initial current draw. This stage of current draw is referred as the mean or maintenance current density, identified as i_{cm}, and reported as A/m^2. The low polarization current is possible because, as the initial current is applied, a calcareous deposit covers the bare steel surface. This deposit primarily consists of calcium carbonates, and its thickness is generally about one-tenth of a millimeter. This deposit acts as barrier and is beneficial for cathodic protection system; it protects the steel from contact with the electrolyte (seawater) and reduces the polarization resistance, thus reducing the demand for more current from anodes. This is the current required density after CP has attained the steady-state protection potential. Typically, the mean potential is more negative than the design protective potential by 0.15–0.20 V. For example, if the design potential was −0.80 V, the required potential at this state would be −0.95 to −1.0 V. This reduced current potential reduces the current demand from the anodes, and it is estimated that the current density is about 50% less than the initial current densities. It may be noted that the demand on current density is affected by the sea current, depth, and temperature. Other geographical factors also have their impact on the current densities. Several tables are available as references for specific areas. Table 6.1 gives some of these details.

At the end of the structures' life, the structure is covered with the calcareous deposits and marine fouling, reducing the current demand to its lowest level. The final current density (i_{cf}) takes this buildup into account, as well as damage to these protective deposits. This final current density is used to determine the appropriate polarization capacity to ensure that the structure remains polarized to a potential of 0.90 to 1.05 V.

This leads to calculations of required anode mass, number of anodes, and current demand for these stages of structures life.

CURRENT DEMAND CALCULATION

The initial current demand I_c (A) is calculated to provide sufficient polarizing capacity and to maintain cathodic protection during the design life. The initial current demand (I_c) is calculated for the individual surface area (A_c), which is in meter^2 for each CP unit. The calculated area is factored into the current density (i_c) (unit is A/m^2) and the coating breakdown factors (f_c) to obtain the initial current demand. The relationship is given in the equation below.

$$I_c = A_c \times f_c \times i_c$$

Similarly, the mean or maintenance current demand I_{cm} and final current demand I_{cf} are also calculated using mean and final surface areas and recorded.

CURRENT DRAIN CALCULATION

Complex offshore structures have various parts and components that are temporary or not considered to be part of the structure, but as long as they are there, these parts will drain the CP current. Mud mats, skirts, piles, well casings, anchors, and mooring chains are some of these current drain sources. All such items should be identified and accounted for as sources of current drain.

The sources of current drain are all to be calculated using similar principles and factors (area, coating breakdown, and current density) to establish drain. For drain calculations for the buried parts of these structures, the current density is taken to be 0.020 A/m^2. The area calculation of open pile ends should include the about five times the diameter of the pile to accommodate the internal area of the pile that is exposed to seawater.

The well-casings are often cemented, which reduces the current drain, thus assuming 5 A per well casing as current drain is normal practice.

The anchor and mooring chains are assumed to be 30 m, and the the sections above seawater level are also accounted for in current drain calculations.

TYPE OF ANODE SELECTION

There are several types of anode that are used for cathodic protection. Standoff, flush-mounted, or bracelet-type are further divided into subtypes based on available weight and attachment methods. These are selected based on the structure itself, the available attachment possibilities, and the size needed to produce current output, and factors to be considered include total anode mass, current drag, and subsea current flow interventions. Anode type selection is also dependent on its utilization factor, which determines the current output.

The selection of anode shape is significantly related to the current output and utilization factor. Long and large anodes of up to 2200 lbs. (1000 kg) are commonly selected for offshore structures; they are often standoff-type and attached to the sections of the structure. The total net anode mass (M_a in kg) is the product of the following relationship. The net anode mass is the mass that is required to maintain cathodic protection for the design life of the structure. The design life (t_r), mean or maintenance current (I_{cm}), and number of hours in a year ($365 \times 24 = 8760$ h) are divided by the anode utilization factor (u) and design electrochemical capacity of the anode (ε), which is measured in ampere-hour per kg (Ah/kg). This relationship is denoted as given below.

$$M_a = (I_{cm} \times t_r \times 8760)/(u \times \varepsilon)$$

The utilization of the anode is the fraction of anode material of a specific design that may be utilized for the design life. Beyond the utilization level of the anode its current output is not reliable. Table 6.7 gives the anode utilization factors for some common types of anode.

Table 6.7 Anode Utilization Factors

The Recommended Anode Utilization Factor (u)	
Type of Anode	**Utilization Factor(u)**
Short flush mounted, bracelet type anodes	0.80
Long Flush mounted anode where L ≥ width and L ≥ thickness	0.85
Long and slender stand-off anode where L ≥ 4r	0.90
Short and slender stand-off anode where L < 4r	0.85

Where; r = radius of cylindrical anode or for non-cylindrical shapes it is taken as c/2π where c is the cross sectional area of the anode shape.

The number of anodes required to achieve required current output is determined for all three stages of design life described below.

The initial polarization stage (I_{cr}) refers to the current demand for the structure's protection during the early life of the CP system, and the current demand at this stage is at its highest level. Once the polarization is established after 4-8 weeks in seawater, the current demand stabilizes, and subsequently, lower current is required to maintain the protection. This lower current occurs during the mean or maintenance stage, which we will discuss next. However, once we have determined the mass of the anode material, we need to determine number of anodes (N) that would be required to meet the current demand (I_c). For this step, we need to determine the type of anode suitable for the structure in question. For offshore structures, depending on the effect of sea current drag or interference, one of any combinations of standoff, flush-mounted, or bracelet-type anodes are selected. This is established by taking into account the total mass required, as well as the space available for installation on the member of the structure. As said above, the size and geometrical shape of the anode also depends on forces such as sea current and the anode's utilization factor. Table 6.7 indicates the utilization factors for different anodes.

When the type and size of the anode is determined, the calculation of the number of anodes (N) is based on following calculations.

The current output (I_c) of an individual anode is the product of the number of anodes and the individual current output (I_a). This is determined using the following relationship between the close circuit potential of the anode (E_a^o) measured in voltage and the design protective voltage (E_c^o), which is $-0.80\,V$, divided by the resistance of the anode (R_a) ohm. This relationship can be written as:

$$I_c = N \times I_a = \left(E_c^o - E_a^o\right)/R_a$$

Given that the difference between the protective voltage and the close circuit potential ($E_c^o - E_a^o$) is the change that determines the current output for the particular CP system, it is also called the driving voltage for the CP design. This equation can be further simplified and written as:

$$\Delta E^o / R_a$$

The above equation is used to calculate current required to keep the structure polarized and protect the structure. As discussed above, the initial protective current (I_{ai}) is greater than the required maintenance and final currents (I_{af}). Similarly, the corresponding anode resistance (R_a) for the initial and final calculations are taken as R_{ai} and R_{af}, respectively. These values are substituted in the equation above, and the process remains the same for each case of calculating current demands.

The typical calculations involved in CP design will include following steps.

- Calculation 1: Determining the anode current output for three current densities

For an impressed-current CP system to protect structures, the selection of an anode involves knowing the outputs of the selected anode types and sizes. It is important that designers make a short list of possible choices, making calculations for each size to optimize the selection process. The calculation of the resistance (R) in ohms is done using the following equation.

$$R = p * (K/L)[\text{In}\{(2L/r) - 1\}]$$

However, the following method based on the Dwight's equation is used more frequently for determining the current output of a selected anode type and size. Dwight's equation works well when the selected anode's length (L) to radius (r) ratio is over a specific number. This ratio is established as $4L/r \geq 16$. When the value of this ratio is <16, other equations, such as the McCoy equation, are often used.

McCoy's equation calculates the resistance (R) in ohms, with the value of R being the product of water resistivity in ohm–cm and the root of the anode area (A) in cm^2. The equation is written as:

$$R = (0.315p)/A^{-0.5}$$

McCoy's equation relies on the principle that the resistance (R in ohms) of an anode in a given electrolyte is equal to the product of the specific resistivity of the electrolyte and factors specific to the shape of the anode.

The equation can be written as:

$$R = p * (K/L)\{ \ln(4L/r)\} - 1$$

where R is the resistance (in ohms) between anode and the electrolyte, L is the length of the anode in centimeters, P is the electrolyte resistivity in ohm–cm (refer Table 6.4), and r is the radius of the anode if it is in cylindrical shape. If the anode is not a cylinder, then r can be calculated as $C/2\pi$, where C is the cross-sectional circumference, K is the constant valued at $0.500/\pi$ or

0.159 if L and r are in centimeters. This constant is 0.0627 if L and r units are in inches.

For determining the current output from an anode, Ohm's law equation $I = E/R$ is then used.

As stated above, the calculation required for determining the cathodic protection to be applied to a structure for its design life must satisfy the structure's initial, maintenance, and final current demands. Assuming that the structure is in the Gulf of Persia, the calculations will be as follows:

To start with, we require the *surface area* of the structure to be protected by cathodic protection. The surface area is calculated from the structure data, including the drawings. To more easily calculate the surface area of a complex structure, engineers commonly divide the structure into suitable and manageable subsections, with the area being calculated for each subsection.

For this calculation, we take a subsection that has the surface area of 18,600 m^2.

In addition to the exposed surface area, we require the *design life* for the CP system or the structure when selecting an anode, and for this example, the design life is 20 years.

We select an aluminum–zinc–mercury alloy anode to address the given surface area and design life. The current E for this type of anode is 0.25 V, and the driving voltage between the anode and an Ag/AgCl reference electrode in seawater is −1.50 V.

The length of the anode is 245 cm.

The weight of an anode is 330 kg.

The anode is not cylindrical, but has a 22 cm by 22 cm square cross section. The core of the anode is 100 mm NPS, Schedule 80 pipe, and the radius r of the core = 5.7 cm.

The circumference r of this anode can then be calculated, where $C/2\pi = (22 + 22 + 22 + 22)/(2^*\pi) = 14.00$ cm.

Aside from its physical dimensions, the anode has a current capacity of 2750 A-h/kg (obtained from anode supplier's data).

So, using all these measures, we can calculate $I = E/R$, as follows.

$$I = 0.25/20 * (0.159/245)[\ln(4 \times 245)/14] - 1$$

Thus, $I = 0.25$ V/0.0410588 ohms = 6.0888 A (approximately 6.10 A)

The next step is the calculation of the number of anodes required to deliver the calculated current.

$$N = (\text{initial current density} \times \text{surface area})/$$
$$(\text{current output per anode} \times 1000\,\text{mA/A}).$$

Substituting the available values from above:

$$N = \left(110\,\text{mA/m}^2 * 18{,}600\,\text{m}^2\right)/(6.1 * 1000\,\text{mA/A})$$
$$= 335.409\,(\text{say}\,336)\,\text{anodes}$$

Next we calculate the maintenance or mean current density. At this stage, we can use the number of anodes required to meet the current density. We also note at this point that the demand for current density will be far less than the initial current required to polarize the structure. The mean current required is 55 mA/m^2, and the calculation for this stage, uses the previous equation for N with the addition of design life as a numerator in the function. Similarly, in the denominator, the anode output is replaced with the current capacity for the selected anode weight. The changed equation appears below:

$$N = (\text{mean current density} \times \text{surface area})$$
$$* (\text{design life in hours})/(\text{current capacity for selected anode weight}$$
$$\times 1000\,\text{mA/A})$$

To convert the design life into hours, we multiply 8760 h in a year by 20, with the answer being 175,200 h.

Substituting the available values from above:

$$N = \left(55\,\text{mA/m}^2 * 18{,}600\,\text{m}^2\right) * (175,\,200)/(2750 * 330 * 1000\,\text{mA/A})$$
$$= 197.498\,(198)\,\text{anodes}$$

Note that this number is lower than the initial number of anodes required.

Finally, we calculate the number of anodes required to provide for the current demands during the final stages of the structure's design life. The calculation is somewhat similar to the N calculation for the initial stage, with one exception. At this stage, the anodes are at the end of their lives, and they have been expanded, losing much of their shape and dimensions. This significant change is, thus, an important factor in the calculations. So, before we continue with the current demand and number of anodes required for final stage, we need to determine the expanded dimension (r) of the anode.

$$r_{\text{expanded}} = r_{\text{initial}} - (r_{\text{initial}} - r_{\text{core}}) \times \text{the anode utilization factor} \ (90\%)$$

By substituting the values from previous calculations for the initial anode requirements above:

$$r_{\text{expanded}} = 14 - (14 - 5.7) * 0.9 = 5.13 \, \text{cm}$$

Now, the final current output per anode is calculated using the Ohms law equation.

$$I = E/R = 0.25 \, \text{V} / \{ 20 * (0.159/245) [\ln(4 * 245)/5.13 - 1] \}$$
$$= 0.25 \, \text{V} / 0.054089708 = 4.62 \, \text{A}$$

The number of anodes required to protect the structure is then calculated using the same equation for N.

$$N = (\text{initial current density} \times \text{surface area})$$
$$/ (\text{current output per anode} \times 1000 \, \text{mA/A}).$$

Substituting the available values from above, with the final current density being 75 mA/m^2, we get:

$$N = (75 \, \text{mA/m}^2 * 18{,}600 \, \text{m}^2) / (4.62 * 1000 \, \text{mA/A})$$
$$= 301.948 \, (\text{say} \, 302) \, \text{anodes}$$

Based on the previous three stage calculations for current demands and numbers of required anodes, we know that the number of anodes required differs significantly between the initial, mean (maintenance), and final stages. Especially the current demand between mean and final and then the difference between initial and final is not much. This is not an acceptable solution. Therefore, we should consider some other types and sizes of anodes if we wish to optimize the anode numbers. Several options should be explored to reach an optimal number that satisfies all three stages, with no large changes from one stage to another in either current output or numbers of anodes.

• Calculation 2: Optimizing the number of anodes

For this exercise, we consider a CP system for the structure with anodes having a different weight and size. As above, the new data are incorporated into calculations for current output and numbers of anodes, as follows. In this new scenario, we also make these changes:

The length of the anode is 225 cm.

The weight of an anode is 250 kg.

The anode is not cylindrical but has a 15 cm by 22 cm square cross-section.

The core of the anode is 75 mm NPS, Schedule 80 pipe. The radius r of the core is 3.75 cm.

The circumference r of this anode can be calculated, where $C/2\pi = (15 + 15 + 22 + 22)/(2*\pi) = 11.77$ cm (approximately 12.00 cm).

The current capacity (CC) of the anode is 2750 A-h/kg (obtained from anode supplier's data).

Calculating $I = E/R$ then leads to:

$$I = 0.25/20 * (0.159/225)\ln[(4 \times 225)/(12) - 1]$$

$I = 0.25$ V$/0.046887$ ohms $= 5.33$ A (approximately 5.5 A)

Now we can calculate the number of anodes required to deliver the calculated current, using the following equation.

$$N = (\text{initial current density} \times \text{surface area})$$
$$/(\text{current output per anode} \times 1000\text{mA/A})$$

Substituting the available values from above:

$$N = \left(110\text{mA/m}^2 * 18{,}600\,\text{m}^2\right)/(5.5 * 1000\text{mA/A}) = 372 \text{ anodes}$$

Next we calculate the maintenance or mean current density. The mean current required is 55 mA/m^2.

$$N = (\text{mean current density} \times \text{surface area})$$
$$* (\text{design life in hours})/(\text{current capacity for selected anode weight}$$
$$\times 1000\text{mA/A})$$

To convert the design life into hours, we multiply 8760 h in a year by 20 to get 175,200 h.

Substituting the available values from above;

$$N = \left(55\text{mA/m}^2 * 18{,}600\,\text{m}^2\right) * (175{,}200)/(2750 * 250 * 1000\text{mA/A})$$
$$= 26.04 (\text{approximately say } 26) \text{ anodes}$$

Finally, we calculate the number of anodes required to provide for the current demands during the final stages of the structure's design life. Before we continue with finding the current demand and number of anodes required for the final stage, we need to determine the expanded dimension (r) of the anode as follows.

$$r_{expanded} = r_{initial} - (r_{initial} - r_{core}) \times \text{the anode utilization factor } (90\%)$$

Case Study 1
Overview of the Structure and the Design Conditions

The system consists of a number of wells in the field that flow to the pump stations. From these pump stations, lines go to four risers that are suspended from four buoyancy cans by a chain. At the top of each of these risers, there is a swivel that attaches to a flexible riser leading to the turret assembly that is housed in the FPSO. In addition, there is another buoyancy can that supports the flexible and rigid risers leading to the subsea export gas line. In total, there are five buoyancy cans with five sets of riser assemblies. There are also four umbilicals leading from the subsea production and pumping area so that two of those umbilicals connect to the turret assembly. The turret is designed to be able to be detached from the FPSO in case of emergencies.

The CP system for the four risers is designed to be isolated from the turret assembly and the FPSO. Each line from the two umbilicals (16 lines in each umbilical) is also isolated from the turret assembly and the FPSO.

A review of the existing system was carried out. The review was initiated due to the failure of one of the links in the chain attached to the rigid riser and a flexible riser attached to a buoyancy can, and the review was intended to identify any role that the CP system may have played in the failure. The terms of the review require focusing on the impressed-current cathodic protection systems (ICCP) and galvanic cathodic protection systems (SCP) to determine if they had some part in causing hydrogen to generate, causing hydrogen-induced cracking of the chain link.

Objective of the Study

The objective of the case study is to review the design of the CP system in reference to the listed international specifications, drawings, and the internal reference documents.
- DNV RP-B401 Cathodic Protection Design
- DNV F 103 Cathodic Protection of Submarine Pipelines by Galvanic Anodes
- ISO 15589-2 Petroleum and Natural Gas Industries—Cathodic Protection of Pipeline Transportation Systems, Part 2: Offshore Pipelines.
- NACE RP-0176 Corrosion Control of Steel Fixed Offshore Platforms Associated with Petroleum Production.
- Supply of Sacrificial Anodes for the Subsea Production System

Continued

- Sacrificial Anode Specification
- Sacrificial Bracelet Anode—Technical Specification
- Sacrificial Bracelet Anode Specification
- Flow line Cathodic Protection Design
- Flow line Cathodic Protection Design
- Protection Design Philosophy for Subsea Production System
- Cathodic Protection Design Basis of Subsea Production System
- Cathodic Protection Design Calculation
- Flow line Cathodic Protection Design.

Points of Reference for the Study

1. Is the CP System designed correctly?
2. Can the design protect the assets throughout the design life?
3. Does the design comply with industry standards?

The CP systems of the various components in the structure were reviewed. The following components were reviewed and investigated by measuring potential readings.

- Recorded the seawater-to-FPSO potential of the CP system in the same water.
- Checked the functioning of all isolation points between the FPSO and umbilicals.
- Checked the use of CRAs in some equipment placed in the seawater with CP systems on nearby structures. Checked if the CP current had developed over potential, possibly damaging the CRAs through hydrogen inhibition on cathodic sites.

CP Design Review

This review is based upon the parameters given for four flow-lines. The pipes are 10.625 in. in diameter, with a 0.869-in. nominal wall thickness. These lines are coated with a 3-in. thick GSPU coating over an FBE base coating. The pipes lay on the floor of the sea (seabed) and are considered unburied. The coating deterioration factors used are in line with ISO 15589-2 for the initial, mean, and final stages of the 25-year design life. The design temperature used is $4\,°C$, which is also the MAST.

The potential criterion used is $-800\ mV_{Ag/AgCl}$, and the current density used is based upon DNV RP B 401 tables 10-1 and 10-2 for tropical climates and depths below 300 m of water. The design has also accounted for temperature adjustments of $0.001\ A/m^2$ per degree over $25\,°C$ temperature. The specified sacrificial Al/In/Zn anodes (Galvotec® CW III or equivalent) are considered to have $-1050\ mV_{Ag/AgCl}$ as specified in DNV RP B 401, table 10-6. In addition, the anodes are considered to have a 90% utilization factor, meeting the table 10-8-requirements of DNV RP B 401.

Potential attenuation calculations are conducted on the basis of a maximum length of 11,361 m, taking in account factors such as protective potential (E_c), driving potential, change in potential (Eme), pipe wall thickness, pipe diameter, resistivity of steel, and mean current density ($i_{cm(a/m2)}$). The design recommendation for the mass and potential attenuation calculation for all flow lines appears to be correct and to meet the requirements of the referenced documents.

CP Design Review of Flow-Lines

The design of the flow-lines is based on the referenced specifications. The cathodic protection design for pipe-in-pipe (PIP) production flow-lines consists of 9.625-in. by 14-in. with lengths of 86,638 ft. and 86,232 ft., and they have been provided with 286 and 265 anodes of 154 lbs. weight each, for a total weight of 44,044 and 40,810 lbs. The anodes are recommended to be spaced at 240 ft. The calculation parameters are in conformance with the specified industry specifications.

CP of Fields Free-Standing Hybrid Risers

The input data is taken from the internal documents. The design is based on the recommendations of DNV RP B 401 and the verification of area calculations, as well as the recommended weight and number of aluminum alloy anodes, and the off-take spool-exterior and internal body structure meets the specification. Similarly, the review of the CP design calculations for the production and gas export FSHR suction pipes shows that the system is meeting specified requirements. The calculation basis and methodology are drawn from the available data and DNV RP B 401 requirements.

Design of the Cathodic Protection for the Riser Acoustic Station

The receptacle is protected with cathodic protection, and it includes an ROV T-bar but does not include a TSA-coated receptacle hanger and instruments, as it is assumed that these are protected by their own CP systems. It is also assumed that the hanger and instruments will not drain any current from the anodes provided for the receptacle.

Calculations of the surface area to be protected by CP account for the area of the receptacle exterior, the Monel metal nut, and the ROV handle. The calculations for the initial, mean, and final current output meet the industry specifications listed above. Flush-mounted Al/In/Zn anodes are recommended to be put on 6-in. tubular supports. The weight calculation is based on the given documents, and it follows DNV RP B 401 procedures.

CP Design Review of the Top Riser Assembly

The design life of the Top Riser Assembly (TRA) is 30 years, considering the recommended use of four bracelet-type anodes of 456 lbs. each. The TRA and gooseneck are both TSA-coated. The utilization factor for the bracelet anodes is taken as 80%, based upon DNV RP B 401, table 10-8. Other parameters, such as anode

Continued

potential, protection potential, anode current capacity, sweater resistivity, and factors used for calculation of anode requirements and weight calculations, are based on DNV RP B 401 and ISO 15589-2.

The design takes into account that the TRA-mounted anodes will drain current to 13 link-tether chains. The tether chains are TSA-coated, and the CP design has considered the tether chains to be 50% bare. The design methodology meets the DNV RP B401 requirements.

Conclusion of the Review and Study

The review of the design, verification of the data, and validation checks from the field were extensively reviewed in reference to the objective set for the study. The question of reference was evaluated from various angles and the following conclusion was drawn. As reviewed from the documents supplied, the components of the CP system conform to NACE and DNV specifications, and other referenced industry standards. A few exceptions to the standards do exist, but those exceptions are deemed to be unique to the structure-related details, and a review of the design calculations indicates that the approach taken is based on sound principles and should not be a cause for concern.

With a few exceptions noted in the body of this report, The CP system is well designed and will protect the structures from external corrosion for the design life.

By substituting the values from the previous calculations for the initial anode requirements, we get:

$$r_{expanded} = 12 - (12 - 5.5) * 0.9 = 4.95 \, cm$$

Now, the final current output per anode is calculated using the Ohm's law equation.

$$I = E/R = 0.25 \, V / \{20 * (0.159/225)[\ln(4 * 225)/4.95 - 1]\}$$
$$= 0.25 \, V / 0.05940236 = 4.208 \, A$$

The number of anodes required to protect the structure is then calculated using the N equation from the initial calculation.

$$N = (\text{initial current density} \times \text{surface area})$$
$$/(\text{current output per anode} \times 1000 \, mA/A)$$

We then substitute the available values, with the final current density being 75 mA/m^2.

$$N = \left(75\,\mathrm{mA/m^2} * 18{,}600\,\mathrm{m^2}\right)/(4.208 * 1000\,\mathrm{mA/A})$$
$$= 331.511\,(\text{approximately say }332)\ \text{anodes}$$

In the second set of calculations, the initial and final anode numbers demonstrate a reasonable difference. However, the anode number for the mean or maintenance current is too low at 26 anodes. So, more optimization calculations should be done to further adjust the numbers.

In optimizing the number of anodes and current output, the accuracy of the length (L) and radius (r) of the anode for both initial and final calculations cannot be overemphasized. Similarly, for the purposes of optimization, designers can consider some variations in the protection potential of -80 V in reference to the standard Ag/AgCl reference electrode in seawater. However, while considering these variations, the adequate protection level must not be compromised.

The above calculation is the most common method used to design cathodic protection for a structure in seawater. Yet, the process tends to give a somewhat conservative estimate of the current output, as well as the number of anodes of a specific shape and size that can provide adequate protection to the exposed steel surfaces of the structure. Given its very conservative approach, the process sometimes tends to produce numbers that may be termed as overdesigned. This shortcoming of the method is addressed through the use of more principles-based calculation. This principles-based approach gives somewhat accurate calculation results based on the determination of the resistance of an anode, the design life, and area of the exposed structure as indicated by the design slope in ohms-cm. The process determines the design life (T) of the structure using following equation.

$$T = (R_a w)/(i_m k S)$$

where T= design life in years; R_a = resistance of single anode to remote seawater (this is similar to $R = p^*(K/L)\{\ln\ (4L/r)\} - 1$ used above, P= water resistivity); S= design slope in ohm-m^2; i_m = the current density for the maintenance; k= the anode consumption rate in kg/A year.

This calculation tool can also be used as a verification tool for the conservative approach discussed above.

The key point in the design of a cathodic protection system is to explore a number of anode types and configurations to optimize the system output and longevity. The use of data from various sources, including anode manufacturers' data, is always helpful. The computerized calculations have made the optimization and current drain calculations much easier to perform and the results are more reliable and quicker. The quality of the anode and the provided data have significant impacts on the accuracy of these calculations and the subsequent effectiveness of the designed CP system. In Section 2 of this book, the quality aspect of anodes and their testing are discussed.

As we have noted in the above discussions, the use of galvanic anodes is more prevalent in CP systems for offshore structures, but the impressed-current CP system is also used in somewhat limited ways. In most cases the impressed-current system is applied by attaching the anode to the structure, with the conductors carried through conduit pipe system to the surface for connection to the current source, which is often a rectifier located on the deck. The lead–platinum or lead–silver anodes are used to produce the currents required to protect the structure. The following case study provides a review of an existing CP system design.

CHAPTER SEVEN

Cathodic Protection of Offshore Pipeline Risers and Associated Equipment

INTRODUCTION

The protection of submarine pipelines from external corrosion is primarily achieved through the combined application of high-efficiency coating and cathodic protection (CP). The CP system renders the exposed steel substrate of the pipeline cathodic in reference to other metallic parts on the substrate. A suitable CP system should be selected for the pipeline, and the selection process may require the designer to choose between a galvanic system and an impressed current system.

The CP design process starts with understanding the lay of the submerged pipeline. The lay of the pipeline includes the topography of the seabed that will support or contain the line, as well as the depth of the pipeline or the riser. As a result, the design activity must begin with an assessment of the environment that will surround the pipeline. This activity is part of the pipeline survey, and we have tried to give some introduction to this most important activity.

Despite the need for continued technological advances, the development of new and more efficient corrosion prevention processes has led to more effective CP methods. More efficient coatings have also improved the lifetime integrity of pipeline systems. Similarly, sophisticated purpose-built pipe-lay vessels have replaced barges in the pipe-laying process. As a result, pipes can be reeled and laid with maximum efficiency. The developments in anode metallurgies and application improvements have added the value to the CP system as well. In the following discussions, some these improvements are briefly considered.

PIPELINE SURVEY AND INSPECTION

A pipeline survey is the collection of data related to an existing pipeline or the conditions of the location where new pipeline may be installed. In both cases the collected data pertains to the electrical potential of the involved structures, including the potential of the existing pipeline. The following description of a pipeline survey addresses both cases.

As technologies have developed more generally, the methods for surveying offshore pipelines have also improved. In the early days CP surveys of offshore pipelines borrowed directly from onshore survey methods, as when the tow-fish trailing wire method was replaced with electronic methods. The accuracy of survey results have improved, in turn, resulting in sound baseline data for the management of pipeline integrity.

Surveys are conducted using remotely operated vessels (ROVs). The ROVs use a three-electrode technique to obtain close-interval offshore potential data. The calibration of the ROV is an important factor in obtaining valid and reliable data. The use of an ROV for a pipeline survey also requires that the CP team determine the data's required level of detail, because the level of detail depends on the intervals between the data collection points. The survey data points can be spaced every 100 m for more accurate data, or the points can be spaced as far as 1 km apart, or even farther. In addition to the accuracy desired, the spacing is also related to the quality of an existing pipeline's coating, or the integrity and potentials of other structures in the electrolyte (seawater) where a new pipeline is being installed.

Short-order changes in the protection levels of coated offshore pipelines are generally predictable. This is a verifiable fact and can be validated through the analysis of past data collected from the field. Given this information, designers can establish a sound baseline for future use, as part of the pipeline integrity plan. The survey process involves the ROV taking potential readings at reasonable intervals. Several pipeline design codes require that current attenuation is modeled during the CP design process as well. Attenuation models predict the potential distribution along a pipeline at various distances from known CP current sources (anodes) attached to the pipeline. Knowing the potential distribution can be useful not only during pipeline design, but also during the lifecycle maintenance of the pipeline. In most cases the minimum potential occurs at the midpoint between two known current source drain points. Attenuation modeling involves several variables that have bearing on the accuracy of the data and eventually the model itself. Some of these key cathodic and anodic variables are listed below. Cathodic Variables:

- Coating condition: size, number, and linear distribution of coating defects (Most models either assume the worst case or a linear distribution of damage at a given coating breakdown percentage).
- Pipeline material and wall thickness
- Pipeline operating temperature
- Pipeline exposure (Most models assume the worst case: pipeline exposed, anodes buried.)

A key anodic variable is the resistance between the anode array and the environment, including connection resistance.

A good model is based on the data input from a logical analysis of the situation. For new pipelines, obtaining reliable data is relatively easy because the data is new and available, and the difference between the pipeline's design and its real status is small. For an existing pipeline, the importance of accurate survey data cannot be overemphasized, especially for attenuation calculations. The subject of attenuation is also discussed in detail in Section 5 of the book.

TYPES OF CP SYSTEM

The use of an external power source in a CP system is called the impressed current process. This can be achieved by creating a small electrical potential in the steel pipe, thus making it a cathode. Cathodic protection can also be achieved by using a more active material that is anodic in reference to the steel. A system that employs the latter method is called a galvanic or sacrificial CP system. Some of the metals used as anodes in these systems are specifically designed to be attached to the pipe as an anode in reference to the exposed steel substrate.

GALVANIC OR SACRIFICIAL ANODE CP SYSTEM

Submerged pipelines are most often protected by a galvanic anode system, as opposed to an impressed current system. The anodes are placed on the pipeline with utmost care to ensure that they are producing the desired current and that they are environmentally safe. Pipelines protected by galvanic anodes must be electrically isolated from other structures, however. So, while designing a CP system for a pipeline, engineers must evaluate the impact of the system on other pipelines within the vicinity of the current

Table 7.1 Evaluating the Case for Galvanic System

Lack of external power source, is one good reason to consider Galvanic anode CP system.	
Galvanic anode system	Requires minimal control and maintenance of equipment during service life of pipeline. Interference problems are almost never felt. Very low possibility of interference current causing problems with adjoining pipelines and structures from galvanic system.

flow from the subject pipeline. The anodes are attached to the pipeline to be protected either individually or in groups, and the current output of the anodes is limited by the driving voltage between the anodes and the pipe. Table 7.1 lists some of the factors that may be evaluated when selecting a pipeline's galvanic anode CP system.

IMPRESSED CURRENT SYSTEM

When an external power source is used in a CP system, that system is called an impressed current system. The impressed current system may use high-silicon cast iron, lead, silver, or graphite anodes connected via a cable to the positive terminal of a direct current (DC) power source. The negative terminal of the DC source is connected to the pipeline to be protected. The power source can be a rectifier or a generator, and sometimes chargeable batteries are also used. The main advantage of an impressed current system is that the current can be varied to meet the requirements of the system, especially when the system's polarized current draw does not need to be as high as its initial draw. Once the system has polarized, the rectifier output can be turned down to the minimum level needed for continued protection. Table 7.2 lists some factors that may influence or limit the selection of an impressed current CP system for a pipeline.

Table 7.2 Evaluating the Case for Impressed Current System

Availability of external power source	To connect to rectifier,
Length of pipe	Practical limitations on anode and rectifier location Insulation resistance of the coated pipeline Longitudinal resistance of pipeline itself.
Suitable in	High resistivity waters, like brackish water, and large estuaries.

OBJECTIVES OF THE CP SYSTEM

The main objectives of a CP system can be briefly listed as follows.

1. The system will provide protection from external corrosion for the design life of the pipeline.
2. The system will accommodate anticipated changes in current requirements over time.
3. The system will provide sufficient current and distribute it to the protected pipeline along its entire length.
4. The system will have anodes able to provide the required current for the design life of the pipeline.
5. The system will be monitored.

FACTORS TO CONSIDER WHEN DESIGNING CP SYSTEMS FOR PIPELINES

Pipelines and SCRs are often protected by a combination of high-efficiency coating and a galvanic anode CP system. The galvanic anode system is established by installing anodes of a selected size and shape at properly spaced intervals. The design ensures that the selected anode type and spacing will provide the necessary current output to meet the current density demands. The current output should satisfy the current requirements for the mean and final stages of the design life of the pipeline as well.

The design calculations should also support the number and spacing of the installed anodes as they relate to the distribution of output current. Often spacing that generally exceeds 1000 ft (about 300 m) must be supported by attenuation calculations to ensure that sufficient current is available to protect the segment of the pipeline between the anodes. Thus, the feasibility of such large spacing must be proven by calculations. We discuss these calculations in subsequent parts of this chapter.

When designing an effective CP system for an offshore pipeline, an engineer must consider several factors. As a result, designers collect copious data and make multiple calculations in order to arrive at the optimal parameters for meeting the objectives of the system. The required data about the pipeline to be protected may include following.

In the subsequent calculations, we note that factors such as coating breakdown and the electrochemical capacities of the anode do not vary during the life of the CP design. In most cases the result of calculations vary only due to the operating conditions. The effect of these differing conditions can be handled by calculating at the extremes of the available data. If the value obtained is small, then choosing the most conservative value could be best option. If, however, the difference is significant, then other variables need to be reevaluated for design calculations. Factors such as temperature change, changes in operating conditions during the life time of pipeline, and coating quality may be factors to revisit in such cases.

COLLECTING DESIGN DATA

Regarding the material of the pipeline, the information to be collected may include:

- Type of material
- Grade of material
- Any special treatment applied to enhance the strength and other properties of the material

Designers must also determine the presence of residual stress in the pipe material that may be abnormal based on the natural conditions of the line pipe. Sources of residual stress may include:

- Stress developed by special fabrication processes
- Method of pipe lay
- Selection of the pipe route
- Lay of the seafloor
- Temperature profile gradient for the length of the pipeline, both in operating and shut-down conditions
- The length of the pipeline to be protected
- Design life of the pipeline
- The wall thickness of the line pipe
- Outside diameter of the pipeline
- Type of coating applied and its thickness
- Mechanical protection applied to the pipeline
- Thermal insulation applied to the pipeline
- Any buoyancy control method used (e.g., concrete coating)
- Pipe exposure and burial or trenching details
- Other structures in the proximity of the pipe route: CP status and historical performance data
- Pipe, fittings, J-tubes, risers, clamps, and other appurtenances
- Construction schedule
- Local laws and regulations

Armed with the above data about the pipeline itself, the CP designer should also obtain data about the environmental electrolyte. For pipelines at depths down to 500 ft, variations in seawater characteristics in the geographical area are not significant. These seawater characteristics include salinity, pH, flow currents, dissolved oxygen, and sea fouling. It is always advised to obtain current data on these characteristics, as well as the performance of existing pipelines and structures in the area.

New discoveries, combined with a high demand for energy, have given much of the aging offshore infrastructure a new lease on life. In particular, offshore pipelines are at the highest risk of catastrophic external corrosion failure if their CP systems are allowed to become ineffective. The lives of offshore assets can be extended, however, through the correct use of survey methods, innovative design software, and low-cost hardware. The CP system designed for such offshore pipelines is described below.

DESIGNING CP SYSTEMS FOR OFFSHORE PIPELINES IN RELATIVELY SHALLOW WATER

When designing a CP system for an offshore pipeline, the corrosion engineer considers several variables that can have serious impacts on the level of protection provided to the pipeline. The variables then lead to decisions regarding the selection of anode material, size, positioning, among other aspects of the system. These variables include the following.

- Design life required (often 20 years)
- Pipe diameter, length, and to-from information
- Geographic location
- Type of coating
- Pipe-lay and installation method
- Water depth
- Burial method
- Product temperature
- Electrical isolation from platforms or other pipelines

The corrosion engineer designing the CP system also considers the method of pipeline installation, as most of the damages to installed anodes and coatings happen at this stage. Suitable allowances in numbers and design corrections may be made account for these possibilities. A considered, conservative approach is advised. However, overdesign is not advised, because it increases the expense of providing CP.

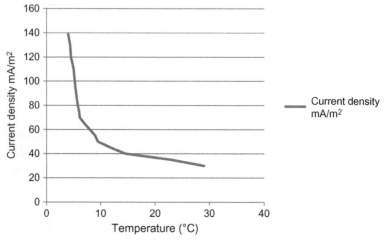

Figure 7.1 Current density graph for sea depths down to 500 ft.

For pipes in sea depths down to 500 ft, the required current density is dependent on the seawater temperature and can be determined using the curve in Figure 7.1.

For pipelines laid deeper than 500 ft, the above specified characteristics can vary significantly as the depth changes. This variation can have a significant impact on the polarization of the CP system and the formation of calcareous deposits. Thus, detailed information should be obtained through the field survey, metrology reports, corrosion test data, and the performance of other pipelines in same area. In addition to these measures, the following data must also be collected and evaluated for effective CP design.

1. Protective current required to meet the established criteria
2. Oxygen concentration at the seabed
3. Electrical resistivity of the electrolyte, with seasonal variations obtained for a number of years. This may include the study of following details:
    (i) Electrical isolation
    (ii) Electrical continuity
    (iii) Deviations from specifications
    (iv) Maintenance and operating data
    (v) External coating and its integrity
4. Pipe burial depth and the extent (length of) of the buried section
5. Length of the exposed section of the pipeline
6. Water temperature at different depths of the sea, including at the seabed
7. Water flow rate and turbulence in different seasons
8. Seabed topography

EXTERNAL CORROSION CONTROL OF DEEP WATER OFFSHORE PIPELINES

Pipelines in deep water face different conditions than the pipelines in the shallow waters discussed above. Thus, deep-water pipelines require a different conceptual approach that translates into obvious specific strategy changes in the design of the CP system. Deep-water pipelines often have special coatings for thermal insulation, and the quality and integrity of the coating system, which has a nearly 99% efficiency level, assumes the primary role in protection. Bracelet-type anodes are still installed on such pipelines, however. Some proprietary designs for standoff anode arrays are also used to protect pipelines in deep water, and some of these arrangements are capable of being monitored from remote locations and ROVs.

As stated earlier, all offshore pipelines are primarily protected from seawater corrosion by coating applied to the steel substrate. This coating system is further supplemented through the application of a CP system. The CP system provides protection at coating defects or "holidays." CP can be applied via impressed current or sacrificial anode systems, and, in deep water, zinc bracelet anodes attached to the pipe are widely used. In recent years, coating systems have developed to include high-performance coatings with efficiencies generally rated at 99%, and these high-performance systems often use fusion-bonded epoxy or multiple layers of different coatings. Aluminum alloy sacrificial anodes are now used in place of zinc anodes as well, and the combination of new coatings and anodes has significantly improved the performance of the protection system.

In order to design a CP system for shallow or deep water, engineers take the following steps, using calculations and modeling. The values of some factors and products used may differ, but the basic principles remain same. The subsequent paragraphs describe these modeling activities.

DETERMINING THE SURFACE AREA TO BE PROTECTED

Once the data are collected and studied, the designer evaluates possible ways to divide the length of the pipeline into sections. This division could be based on current density demand, which is often affected by the depth, operating temperature, or burial conditions of the pipe's sections. Once the sections are determined, the surface area to be protected can be calculated for each section. Within each section of the pipeline, surfaces to be protected include risers, tie-in spools, expansion loops, valves, and tees.

DETERMINING THE CURRENT DEMAND

After determining the total surface area to be protected, the current demand for each section is calculated, using following equation:

$$I_c = A_c f_c i_c \qquad (7.1)$$

where I_c is the current demand in amperes for the section of pipe, A_c is the total surface area calculated for the section of pipeline, f_c is the coating break-down factor from Tables 6.5 and 6.6, i_c is the current density per square meter (See Table 6.1 and Figure 7.1 for pipes under 500 ft depth). Once the current demand (I_c) is established, the next step in the calculation should be to determine the anode type and dimension. This determination will lead to the calculation of anode mass.

SELECTION OF ANODE TYPE

The selection of the type of anode is dependent on factors such as fabrication, installation, and maintenance ease. The anode shape suggests the type of equation to use for calculating the anode's resistance. Similarly, the core of the anode dictates the anode utilization factors to be inserted in the equation. For pipelines, bracelet-type anodes are most common. These anodes are mounted flush with the concrete coating. As a result, the thickness of the concrete coating affects the dimensions of the selected anode, which should be optimized to provide a current output that meets the designed current demand. Most offshore pipelines use aluminum alloy bracelet anodes. These anodes come in two basic shapes, square-shouldered and tapered. These shapes have specific purpose in the design of the pipeline.

- The square-shouldered anodes are typically used on pipes that are coated with concrete as a buoyancy control. The anode with square shoulders is attached in the recess of the coating, so that it is just below the top surface of the concrete. Bracelet anodes are attached in this way to protect the anode during the pipe-laying process.
- The tapered anodes are designed to be installed on pipelines that are coated with FBE or similar types of thin-film coating. The anodes on FBE-coated pipes are tapered only at the ends to allow for easy transition over the stringers of reel during the lay process.

Anodes are particularly at risk from mechanical damage when the pipeline travels over the stinger on the back of the lay barge. Even with these tapered designs, non-weight-coated pipelines still sustain anode damage, which can in turn cause coating damage. Several methods are used to combat this problem. For example, cast-on polyurethane tapers are gaining popularity, and mounting both halves of the bracelet on top of the pipe is also a common technique when the pipe is laid from a reel barge and the anodes are attached offshore.

ANODE MASS CALCULATION

The total anode mass dictates the number of anodes to be used, as well as the anode distribution required to provide protective current to the pipeline. The total mass of the anodes is calculated based on the mean current required to maintain protection throughout the design life of the pipeline. These calculations are carried out separately for each section of the pipeline to be protected. The mass calculation is done for the design life of the pipeline, using the following relationship between mean current (Equation (7.2)), design life in years, electrochemical capacity (ε) of the anode, and anode utilization factor (u) as given in Table 6.7. In the equation below, the design life in years is converted to hours by multiplying the number of years by the number of hours in a year, 8760.

$$m = I_{cm}t_{dl}(8760/u\varepsilon) \qquad (7.2)$$

where m is the total net anode mass required for the pipeline section, the anode mass is in kg, I_{cm} is the mean current demand in amperes, t_{dl} is the design life in years, u is the anode utilization factor, and ε is the electrochemical capacity of the anode measured in ampere hours per kg (Table 7.3).

When calculating the total anode mass, the anode's electrochemical capacity is often taken from reported anodic current density values that are in excess of 1000 mA/m^2, and this is generally an acceptable value for a system operating at temperatures not exceeding 50 °C. For temperature outside this range or in cases in which the conditions will change over the design life of the pipeline, the use of one value for anode electrochemical current capacity is not appropriate. In such conditions the equation for the calculation of anode mass should be suitably modified as given below. The values of i_c, f_c, and ε are not constant but vary over the lifetime of the pipeline.

$$m = (A_c 8760)/u \int_t i_c f_c (1/\varepsilon) d_t$$

Table 7.3 Design Values for Galvanic Anodes

Anode Type	Anode Surface Temperature °C	Immersed in Seawater		Buried in Seawater Sediments	
	Interpolation of Intermediary Temperature is Permitted	Potential Ag/ AgCl/ Seawater	Electrochemical Capacity(ε)	Potential Ag/ AgCl/Sea water	Electrochemical Capacity(ε) A-h/kg
		mV	A-h/kg	mV	A-h/kg
Aluminum	≤ 30	-1050	2500	-1000	2000
	60	-1050	2000	-1000	850
	80*	-1000	900	-1000	400
Zinc	≤ 30	-1030	780	-980	750
	> 30 to 50**			-980	580

*The Aluminum anode surface temperature should not exceed 80 °C
**The Zinc anode surface temperature should not exceed 50 °C
For non buried pipeline the anode surface temperature should be taken as pipeline external temperature.

CALCULATION OF NUMBER OF ANODES REQUIRED

Once the total anode mass is determined using the equation discussed above, the designer can determine the number of anodes required to meet the mean and final current demands of the pipeline. The final dimensions and net mass of an individual anode is optimized to get the best results. This is achieved by executing a number of reiterative calculations using the following equation. The use of a computer program, such as Microsoft Office Excel Sheet® or MATLAB®, are best suited for this type of reiterative calculations. However, a number of other software products are now available to do these optimizing calculations. The relevant equation is:

$$n = m/m_a \qquad (7.3)$$

where m is the total net anode mass in kg, n is the number of anodes to be installed, and m_a is the individual anode mass.

This calculation now leads us to the distribution (spacing) of anodes along the length of the pipeline to provide CP. The maximum spacing of anodes depends on the number of anodes required. In practical application, bracelet anodes used for CP of pipes are set on a number of weld joints. The sled type anodes are based on the calculated distances, however. Anode spacing can be validated and optimized by calculating the required end-of-life current output (I_f) of an individual anode in the system, which is derived from the relationship of the total current required (I_{cf}) of each anode (n). This relationship is expressed as follows.

$$I_f = I_{cf}/n \qquad (7.4)$$

In most cases the required end-of-life current output (I_f) differs from the actual end-of-life current output (I_{af}). This difference occurs because, in the calculation of required current output, factors such as the driving voltage and circuit resistance (R_a) are not considered. The driving voltage is the potential difference: the difference of the design protection potential (E_c) and the anodes' closed-circuit potential (E_a), measured volts. The potential difference can be calculated using following equation.

$$I_{af} = (E_c - E_a)/R_a \qquad (7.5)$$

For all practical purposes, the actual end-of-life current output (I_{af}) should be considered to be equal to or greater than the required end-of-life current output (I_f). In Equation (7.5) we have used the denominator R_a as the circuit resistance, which is practically the anode resistance in ohms. This value can be calculated for different types of anodes. The factors that affect the values are the environmental resistivity (ρ) and the geometry of the anode.

The equation used for bracelet-type anodes is simple because the anode's geometry is simple.

$$R_a = 0.315\{\rho/(A^{0.5})\} \qquad (7.6)$$

For sled-type slender anodes, the calculation includes the length and radius of the anode, as the following equation indicates.

$$R_a = \rho/(2\pi L)[\ln(4L/r) - 1] \qquad (7.6a)$$

where R_a is the anode resistance in ohms, which is equivalent to the circuit resistance; p is the environmental resistivity; (for resistivity of seawater, see Table 6.4 and Figure 7.2); L is the length of the anode in meters; and r is the

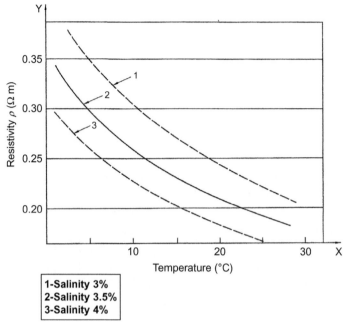

Figure 7.2 Seawater resistivity as function of temperature.

radius of the anode in meters, which is the ratio of the cross-sectional perimeter to 2π, or $r = C/2\pi$.

The above calculation of resistivity applies to pipelines in seawater. If the pipe is exposed to the seabed, then the actual resistivity should be obtained by measurement or from previous data. A default value of 1.5 ohm meters could be used only if the actual data cannot be obtained.

CHECKING FOR THE VALIDATION OF CP DESIGN

The above-discussed methodology for designing a CP system is completed, and the anode type, number, and distribution are finalized. The design can now be checked for functionality. The validation process indicates that the distribution of the pipe-to-electrolyte potential and the distance from the anode, also called the drain point, to the pipeline are adequate. This validation process is called attenuation calculation. We discuss attenuation of protection calculations in the next section.

ATTENUATION OF PROTECTION

Validating the design calculations requires the engineer to consider the factors listed below. In this discussion the term drain point refers to the anode location, or the anode from which the current is emanating and traveling to the cathode (the pipe). All potentials are measured and reported in volts.

- Pipe-to-electrolyte potential at the drain point
- Pipe-to-electrolyte potential at a set distance x from drain point
- Pipe-to-electrolyte potential at the midpoint of the interval between two adjacent anodes
- The flow of current to the pipe measured in amperes
- Current flowing into the pipe at a distance x from the drain point
- Current flowing into the pipe at the midpoint of the interval between two adjacent anodes
- Half of the distance between two adjacent drain points in meters.
- The attenuation constant of the pipeline section (α) in reciprocal meters. (This value can be calculated as the square root of the ratio of Linear electric resistance (per meter) (R_L) and the leakage or transverse resistance (R_t) (ohm-meters), which is expressed as $(R_L/R_t)^{0.5}$.)
- Resistance of the section of pipeline (ohms) (This value is calculated using the relation $(R_L - R_t)^{0.5}$, where R_L is the linear electric resistance of the section of pipeline calculated as p/A_W, with p being the specific resistance of the pipeline material (ohm-meters) and A_W being the cross-sectional area of the pipe wall.

R_t is the leakage or the transverse resistance obtained using the $R_0/(\pi D_o)$, where D_o is the external diameter of the pipe and R_0 is the pipe-to-electrolyte insulation resistance, measured in ohms square meters. This value can be obtained from the field and is based on the following:

- The type of coating and its damage, if any
- Exposure to the environment (seawater, seabed, sediments)
- Design life of the pipeline
- Method of pipeline installation

Figure 7.3 illustrates a typical drain point, with the points and notations used as factors in attenuation calculations. The half-distance between two adjacent drain points, in meters, is annotated with the letter L, and x is the pipe-to-electrolyte potential at a set distance. The attenuation calculations are carried out to derive the values of the pipe-to-electrolyte potential at the drain point (E_x).

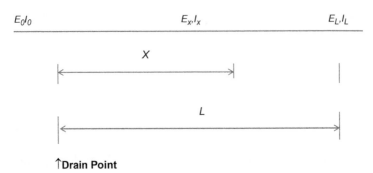

Figure 7.3 Drain point.

Where:

$$E_x = E_0 \cosh \alpha x - R_k I_0 \sinh \alpha x$$
$$E_0 = E_x \cosh \alpha x - R_k I_x \sinh \alpha x$$
$$I_x = I_0 \cosh \alpha x - (E_0/R_k) \sinh \alpha x$$
$$I_0 = I_x \cosh \alpha x - (E_x/R_k) \sinh \alpha x$$

A typical pipeline segment with a uniformly spaced ($2L$) number of anodes serving as the drain points will have the potential (E_x) and current (I_x) flowing at any distance (x). The pipe-to-electrolyte potential is then calculated using following equation.

$$E_x = E_0\{(\cosh L - x)/(\cosh \alpha L)\}$$

And the current can be determined using the following relationship.

$$I_x = I_0\{\sinh \alpha (L - x)/(\sinh \alpha L)\}$$

The drain point calculations are used to validate the design of the CP system, given the number of anodes and their distribution derived from the design calculations discussed earlier in the chapter. The above discussions and sample calculations describe manual calculations. As stated earlier, the majority of these calculations can be done using software that has better calculation abilities and can accomplish a wide variety of calculations if the correct data is entered. The following is the extract of one such calculation done for various structures in seawater, including risers. These calculations are based on the model developed by Uhlig and Revie and used in DNV RP F 103.

SCR CATHODIC PROTECTION SAMPLE CALCULATIONS
Cathodic Protection Analysis Program—SCR

Input of Initial Parameters

Resistivity of Carbon Steel (ohm-m), DNV RP F103, 5.6.10	$\rho Fe := 0.2 \times 10^{-6} \, \Omega \, m$
Design life	$t_d := 25$ years
Design Life Safety Factor	$SF_{td} := 1.0$
Diameter of Flowline (in)	$d_{fl} := 8.625 \cdot in$
Wall Thickness of Flowline (ft)	$wt := 1.181$ in
Length between Sources (ft)	Length := 19,873 ft
Corrosion Potential-Volts (vs. Ag/AgCl/ Seawater)	$EC := 0.685$ volt
Potential at Non-zero Source 1, i.e., at SCR Hull Piping-Volts (vs. Ag/AgCl/Seawater)	$E1 := 1.05$ volt
Anode Potential at Non-zero source 2-Volts (vs. Ag/AgCl/Seawater)	$EA := 1.05$ v
Joint length	$Jt := 40$ ft
Field Joint length	$FJt := 2.5$ ft

Source Anode Selection

Number of Anode Source Locations	`1 ▼`
Type of anodes at Source 1 (Bracelet/Standoff)	`Define Resistance ▼`
Number of anodes at Source 1	$N_1 := 0$
Environmental resistivity at Source 1	$\rho_{S1} := 0.360 \, \Omega \, m$
Type of anodes at Source 2 (Bracelet/Standoff)	
`Standoff ▼` Number of anodes at Source 2	$N_2 := 3$
Environmental resistivity at Source 2	$\rho_{S2} := 0.360 \, \Omega \, m$

Define Resistance at Source (if anode is not selected above)

User defined resistance at Source 1	$R_{D1} := 1000 \, \Omega$
Mass of user defined resistance at Source 1	$M_{D1} := 0$ lb
User defined resistance at Source 2	$R_{D2} := 100 \, \Omega$
Mass of user defined resistance at Source 2	$M_{D2} := 0$ lb

DEFINE ANODE PROPERTIES
Bracelet Anode Properties

Bracelet Anode length	$La_B := 16$ in
Bracelet Anode taper length	$Lt_B := 0$ in

Bracelet Anode thickness $t_B := 1.5$ in
Bracelet Anode weight $Ma_B := 82.2$ lb
Gap between bracelet anodes (assumed) $Gap := 24$ in
Anode Utilization Factor $UF_B := 0.80$

Define Standoff Anode Properties

Standoff Anode length $La_{SO} := 8$ ft
Standoff Anode thickness $t_{SO} := 6$ in
Standoff Anode width $w_{SO} := 6$ in
Standoff Anode weight $Ma_{SO} := 325$ lb
Standoff Anode utilization factor $UF_{SO} := 0.90$
Standoff Anode core radius $Rc_{SO} := 1.00$ in
Standoff Anode Resistance Correction Factor $f_{cor} := 1.3$
    B401 Table 6.7.1 (Note 1)

Define Hull Mounted (Flush) Anode Properties

Hull Anode total length (including mounts) $LTa_H := 29$ in
Hull Anode length $La_H := 24$ in
Hull Anode thickness $t_H := 2.5$ in
Hull Anode width $w_H := 5$ in
Hull Anode weight $Ma_H := 29$ lb
Hull Anode utilization factor $UF_H := 0.85$
Hull Anode core thickness $tc_H := 0.25$ in

Define Properties when not using anodes

Utilization factor for defined resistance (no $UF_D := 0.80$
    defined anode)

COATING BREAKDOWN FACTORS

Coating Breakdown Factor | ISO   ▼ |

ISO—Insulation Material | Thermal   ▼ |
Calculated Coating Breakdown Factor $f_{cf} = 0.45\%$
SOURCE 2—RES ISTANCE
      CALCULATION
Calculated Resistance at Source 1 $RA1 = 1 \times 10^3 \ \Omega$
Calculated Resistance at Source 2 $RA2 = 0.046 \ \Omega$
Temperature of Internal Product (F) $Temp := 300$
DNV—Maximum Internal Product $Temp_{LimitC} = 309.6$
    Temperature for Coating (F)
DNV—Maximum Internal Product for Field $Temp_{LimitFJC} = 309.6$
    Joints (F)
Exposure Condition (Non–Buried/Buried)

| Non-Buried ▼ |
$CD = 0.1 (A/m^2)$

Protection Mean Current Density (DNV RP
F103, Table 5-1)

Design Protective Potential (Volts) $EP := 0.8$ volt

Polarization by Protection Current Density or $P := EP - EC, P = 0.115$ V
Potential Shift (Volts)

Current Capacity at Source 1 (Ah/lb) $CC1 := 0.8 \times 2500$ amp h kg^{-1}

Current Capacity at Source 2 (Ah/lb) $CC2 := 0.8 \times 2500$ amp h kg^{-1}

Rate of consumption (lbs/amp.year) from $Rate := 8.4 . \frac{lb}{amp.year}$
Table 1.B of Ref. [3]

Current Required to Protect Resistive Source 1 $SC1 := 0.0$ amp
(Amps: if not applicable use zero)

Current Required to Protect Resistive Source 2 $SC2 := 0.0$ amp
(Amps: if not applicable use zero)

Voltage Profile

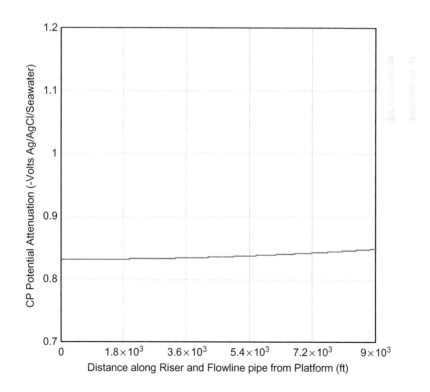

CP Potential Attenuation (-Volts Ag/AgCl/Seawater)

Distance along Riser and Flowline pipe from Platform (ft)

Current Profile

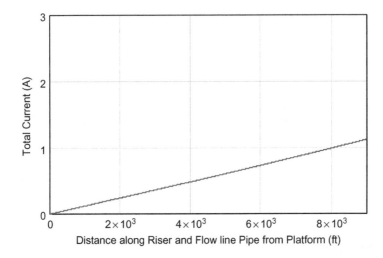

Distance along Riser and Flow line Pipe from Platform (ft)

Results Summary

Voltage Check	$\text{Check}_{\text{Voltage}} = \text{``PASS''}$	
Life Expectancy Check	$\text{Check}_{\text{LifeExpect}} = \text{``PASS''}$	
Mass Check	$\text{Check}_{\text{Mass}} = \text{``PASS''}$	
Coating Breakdown Factor	$\text{CBF} = \text{``ISO''}$	$f_{cf} = 0.45\%$
Distance to Low Point Potential (ft)	$D_{\text{LPP}} = 2 \text{ ft}$	
Low Point Potential (-Volts AgCl)	$V_{\text{LPP}} = 0.831 \text{ V}$	
Potential at $x = 0$ (-Volts AgCl)	$V_{\text{S1}} = 0.831 \text{ V}$	
Potential at $x = L$ (-Volts AgCl)	$V_{\text{S2}} = 0.92 \text{ V}$	
Current Required of Source 1 ($x = 0$)	$C_{\text{S1}} = 2.187 \times 10^{-4} \text{ A}$	
Current Required of Source 2 ($x = \text{Length}$)	$C_{\text{S2}} = 2.852 \text{ A}$	
Total Current	$\text{TC} = 2.852 \text{ A}$	
Envelope IR Drop at $x = L$ (Volts)	$V_{\text{drop}} = 0.13 \text{ V}$	
Mass Required at Source 1	$\text{Mass } 1 = 0.066 \text{ lb}$	
Mass Required at Source 2	$\text{Mass } 2 = 765.47 \text{ lb}$	
Total Mass Required	$\text{Mass T} = 765.536 \text{ lb}$	
Number and Resistance of Anodes at Source 1	$N_1 = 0$	$\text{RA1} = 1 \times 10^3 \ \Omega$
Number and Resistance of Anodes at Source 2	$N_2 = 3$	$\text{RA2} = 0.046 \ \Omega$
Design Life	$t_d = 25 \text{ years}$	$\text{SF}_{td} = 1$
Life Expectancy (Years)	$\text{LifeExpect} = 25.868$	

Minimum required Anode Mass (lbs)	MinAnode Mass = 942.3
Anode Mass Provided (lbs)	Anode Mass$_T$ = 975
Minimum required Anode Mass at Each Location (lbs)	Anode_Loc = 942.277

FLOWLINE CATHODIC PROTECTION SAMPLE CALCULATIONS
Cathodic Protection Analysis Program Flow Line
Project: Flow Line (PLET to ILS)

Input of Initial Parameters

Resistivity of Carbon Steel (ohm-m), DNV RP F103, 5.6.10	$\rho Fe := 0.2 \times 10^{-6}\ \Omega\,m$
Design life	$t_d := 25$ years
Design Life Safety Factor	$SF_{td} := 1.00$
Diameter of Flowline (in)	$d_{fl} := 8.625$ in
Wall Thickness of Flowline (ft)	$wt := 1.181$ in
Length Between Sources (ft)	Length := 25,374 ft
Corrosion Potential-Volts (vs. Ag/AgCl/Seawater)	EC := 0.685 V
Potential at Non-zero Source 1, i.e., at SCR Hull Piping-Volts (vs. Ag/AgCl/Seawater)	E1 := 1.05 V
Anode Potential at Non-zero source 2-Volts (vs. Ag/AgCl/Seawater)	EA := 1.05 V
Joint length	Jt := 40 ft
Field Joint length	FJt := 2.5 ft

Source Anode Selection

Number of Anode Source Locations

Type of anodes at Source 1 (Bracelet/Standoff)

| Standoff ▼ Number of anodes at Source 1 | $N_1 := 2$ |
| Environmental resistivity at Source 1 | $\rho_{S1} := 0.360\ \Omega\,m$ |

Type of anodes at Source 2 (Bracelet/Standoff)

Standoff ▼ Number of anodes at Source 2	$N_2 := 2$
Environmental resistivity at Source 2	$\rho_{S2} := 0.360\ \Omega\,m$
Define Resistance at Source (if anode is not selected above)	
User defined resistance at Source 1	$R_{D1} := 100\ \Omega$
Mass of user defined resistance at Source 1	$M_{D1} := 0$ lb
User defined resistance at Source 2	$R_{D2} := 100$ ohm
Mass of user defined resistance at Source 2	$M_{D2} := 0$ lb

COATING BREAKDOWN FACTORS

Select Coating Breakdown Factor (ISO /DNV/
Defined)

$\boxed{\text{ISO} \quad \blacktriangledown}$

Define Coating Breakdown Factor (if not ISO or
DNV)

$f_{cfD} := 1.00\%$

Calculated Coding Breakdown Factor

$f_{cf} = 0.45\%$

Calculated Resistance at Source 1

$RA1 = 0.068 \; \Omega$

Calculated Resistance at Source 2

$RA2 = 0.068 \; \Omega$

Temperature of Internal Product (F)

$Temp := 300$

Exposure Condition (Non-Buried/Buried)

Protection Mean Current Density (DNV RP F103,
Table 5-1)

$CD = 0.1 (A/m^2)$
$EP := 0.8 \; V$

Design Protective Potential (Volts)

Polarization by Protection Current Density
or Potential Shift (Volts)

$P := EP - EC,$
$P = 0.115 \; V$

Current Capacity at Source 1 (Ah/lb)

$CC1 := 2000 \; \text{amp h kg}^{-1}$

Current Capacity at Source 2 (Ah/lb)

$CC2 := 2000 \; \text{amp h kg}^{-1}$

Rate of consumption (lbs/amp.year)
from Table 1.B of Ref. [3]

$Rate := 8.4 \frac{lb}{amp.year}$

Current Required to Protect Resistive Source 1
(Amps: if not applicable use zero)

$SC1 := 0.0 \; \text{amp}$

Current Required to Protect Resistive Source 2
(Amps: if not applicable use zero)

$SC2 := 0.0 \; \text{amp}$

Current Profile

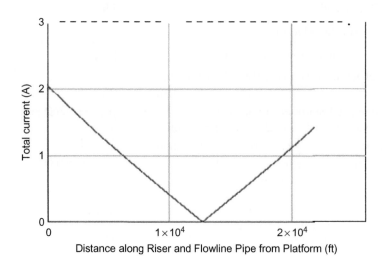

Results Summary

Voltage Check	Check$_{\text{Voltage}}$ = "PASS"	
Life Expectancy Check	Check$_{\text{LifeExpect}}$ = "PASS"	
Mass Check	Check$_{\text{Mass}}$ = "PASS"	
Coating Breakdown Factor	CBF = "ISO"	f_{cf} = 0.45%
Distance to Low Point Potential (ft)	D_{LPP} = 12,687 ft	
Low Point Potential (-Volts AgCl)	V_{LPP} = 0.867 V	
Potential at $x = 0$ (-Volts AgCl)	V_{S1} = 0.91 V	
Potential at $x = L$ (-Volts AgCl)	V_{S2} = 0.91 V	
Current Required of Source 1 ($x = 0$)	C_{S1} = 2.046 A	
Current Required of Source 2 ($x = $ Length)	C_{S2} = 2.046 A	
Total Current	TC = 4.093 A	
Envelope IR Drop at $x = L$ (Volts)	V_{drop} = 0.14 V	
Mass Required at Source 1	Mass 1 = 549.236 lb	
Mass Required at Source 2	Mass 2 = 549.236 lb	
Total Mass Required	Mass T = 1098.472 lb	
Number and Resistance of Anodes at Source 1	N_1 = 2	RA1 = 0.068 Ω
Number and Resistance of Anodes at Source 2	N_2 = 2	RA2 = 0.068 Ω
Design Life	t_d = 25 years	
Life Expectancy (Years)	LifeExpect = 27.013	
Minimum required Anode Mass (lbs)	MinAnodeMass = 1203.1	
Anode Mass Provided (lbs)	Anode_Mass$_T$ = 1300	
Minimum required Anode Mass at Each Location (lbs)	Anode_Loc = 601.553	

Coating and High-Efficiency Coating Systems

CHAPTER EIGHT

Coating for Corrosion Prevention

INTRODUCTION

Protection from corrosion can be achieved using different methods, both alone and in combination. As a result, operators often combine methods based on their complimentary properties. This type of multilayered corrosion protection system includes different groups of corrosion protection methods, such as the application of coating in conjunction with cathodic protection.

Different types of coatings are applied to achieve different objectives. For example, specialized coatings are applied to protect pipes from abrasion damages during the pipe-laying process or from the surface conditions of the lay. Coatings are also developed to protect metal structures in different temperature ranges or to enhance the flow efficiency of the pipe's internal surfaces. In the latter case, the flow efficiency is increased by reducing the coefficient of friction between the pipe's surface and the fluid being transported.

Corrosion protection coatings are also designed to address various conditions. Some of these coatings are simple single-layer coatings applied to the steel substrate, while multilayered coatings provide more robust protection in different service conditions.

However, different layers of protection can exist within a coating group. Such a combination of different coating layers results in a very effective system that provides protection from a wide range of corrosion mechanisms. For example, a combined layer system could provide protection from mechanical damages and electrochemical reactions causing corrosion damage.

HISTORICAL USE OF COATINGS TO PROTECT MATERIALS

Throughout history, various natural products have been used to protect materials of value to humans. Long ago, people used animal fat and

Corrosion Control for Offshore Structures
115

animal fat mixed with natural pigments to coat treasured items, and later they replaced animals fat with vegetable oils to improve protection. In both cases, color pigments seem to have been used for identification or aesthetics.

These developments led to the creation of paints, which have been used for ages to protect material. In fact, the development of paint dates back to prehistoric periods, when people coated objects with resins and pigments extracted from natural resources. However, more serious developments in coating technology can be attributed to the application of zinc, followed by the use of zinc oxide in the early nineteenth century. Since then, surface treatments have evolved from inorganic zinc coating through zinc primers to the current protective coatings. The development of polymers and film-forming coatings occurred early in this process. The epoxy resins, which were mostly amine cured, subsequently emerged as the main basis of coating systems, and over 75 different categories of epoxy coatings are now available.

Inorganic zinc coating systems have also improved from water-based products to the current solvent-based ethyl silicate self-curing coatings that compete with contemporary coating systems using polyurethane.

Early coating systems had high levels of volatile organic contents (VOC); however, a growing awareness of the environmental and health risks associated with VOC led to restrictions introduced in the Clean Air Act of 1978.

These restrictions have improved the epoxy system, making it usable in more demanding conditions. In that time, inorganic zinc products have similarly improved in their application and use.

The other line of coatings systems that has developed since 1978 is the aliphatic polyurethane topcoats.

At this point, it is essential to identify a few terms commonly used in the coating field. Some of these terms are often used interchangeably when speaking of coating. The terms in question are "painting" and "coating." Understanding these two terms is especially important because we are preparing to consider the main subject of this chapter, the high-efficiency coating system. As we will discuss further, early surface applications were referred to as paint, and later applications were referred to as coatings. Yet, an obvious question arises: What is the difference?

This chapter attempts to provide more practical definitions for both terms, though making distinctions between the two applications can be very difficult.

DEFINITION OF PAINT

The liquid in traditional paint is oil with possible additions of natural resins and pigments. This oil-based combination is applied to a substrate, and it reacts with oxygen in the air to form a solid film that provides the protection to the substrate.

The reaction with oxygen is active throughout the life of the paint, leading to its deterioration over the time. We may conclude that paints provide less permanent protection against elements and thus its limitation to prevent corrosion for longer life of the material.

The modern paint has an aqueous base and may contain either latex, polyvinyl acetate, or acrylic. On application, the water-based paint coalesces into a solid film to provide weather resistance, and if pigments and plasticizers are added, the paint also adds aesthetic value.

DEFINITION OF COATING

The term coating points to a chemical compound, and a coating is essentially composed of synthetic resins or inorganic silicate polymers. When applied to a prepared surface, these chemicals will form a coating that resists the harsher environments of industry and sea. In terms of adhesion, toughness, and resistance to weather and seawater, these applications perform far better than a paint system.

CORROSION PROTECTION MEASURES

Corrosion protection measures are divided into active and passive processes. Electrochemical corrosion protection plays an active part in the corrosion process by changing the substrate's electrochemical properties and, as a result, its reactions to the environment. This shift in interactions includes changes to the electrochemical potential of the involved materials.

As stated earlier, the application of coating is intended to protect the substrate from reacting with its environment, so that it will not engage in an electrochemical process and corrode. Coatings provide protection moisture, dissolved gases, acids, and other reactants in the environment. Other key objectives of applying a coating are listed below.

- Resistance from physical, chemical, and biological degradation
- Dielectric insulation
- Thermal insulation (provided by some coatings)

Now we know that the objective of coating is to provide a barrier between the corrosive environment and the metal substrate to be protected. The required barrier can be achieved using various means, including coating the substrate with electrically insulating material or changing the electrochemical behavior of the protected material by changing the flow of ions so that the material to be protected is more cathodic to its environment. However, in most applications, both methods are simultaneously applied to support each other in protecting the metal from corroding. This combination allows the one method to cover the deficiencies of the other method.

In spite of very efficient coating application systems, no coating system can be treated as defect-free. The coating industry often claims to have a 100% efficient system, and they are correct regarding the efficiency of their product. However, in practical terms, efficiency is not always so high, mainly due to construction and transportation stresses that damage the applied coatings. This leaves possibility of defects in the coating called holidays. Holidays may be small pores, and they become anodic in the corrosion process. Due to the relatively small size of these holidays, a small anode to a large cathode ratio is created, and the metal around the holidays tends to corrode rapidly and fail the structure. Figure 8.1 graphically illustrates the general reaction that takes place on coated steel, including the permeation, holiday, coating disbonding, and electrochemical cell formation. These

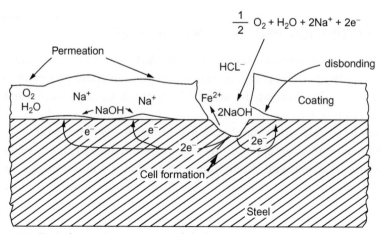

Figure 8.1 Typical corrosion reaction at failed coating.

deficiencies in coating are covered by electrochemical corrosion protection methods, such as cathodic protection, which is discussed in the Section 3 of this book.

There are various types of coating that have been developed to address specific environment and service conditions. In a very broad sense, they can be classified into the following groups according to how they are made.

- Organic coating
- Enamel
- Metallic coating
- Cement mortar

These coatings have different properties, and they are suitable for application to specific environments. Matching the coating properties to the environment is the right way to select suitable coating. In fact, a coating's specific properties are the reason for selecting it for a specific application, and coating properties should remain stable and not change in the given environment for a significantly long period, often exceeding the designed life of the structure they are applied to as protection.

The properties of a coating must remain constant under service conditions for long-term corrosion protection. The stability of coating material in the service environment includes the coating materials' chemical and aging stability against heat and ultraviolet (UV) irradiation. Depending on the type and ultimate objective of the specific coating, it can be applied in a relatively thick layer measured in millimeters or in thin layers measured in microns or mils. In the thin layer coating system, the protective action is from the pigments in the primer that inhibit corrosion. In due course of time, this property of the coating diminishes as the corrosive environment reacts. Coatings that are exposed to electrolytes are designed according to different principles; they are based greater thickness and low permeation rate. This allows for the protection of the substrate even when it is no longer passive or inhibited.

ORGANIC COATING

Organic coatings show some degree of solubility and permeability in regards to corrosive media. These properties of coatings are also described as the permeability and ionic conductivity of the coating system. There are several inspection and test methods that determine the acceptable limits of permeability. Organic coatings can dissolve oxygen and water molecules and allow them to penetrate the entire coating. The corrosion reaction that ensues is best described by the following equation:

$$4Fe + 3O_2 + 2H_2O = 4FeOOH \qquad (8.1)$$

Further action occurs when FeOOH separates the H_2O, and this separation leads the to the corrosion process in which permeation of O_2 occurs. The water corrosion is negated by the action of H_2 as described in the following reaction:

$$2H_2O + 2e^- = 2OH^- + 2H \rightarrow 2OH^- + H_2 \qquad (8.2)$$

Oxygen penetration is further defined by the following reaction:

$$jv/Lm^{-2}h^{-1} = 3.6 \times 10^5 \left\{ P/cm^2 s^{-1} bar^{-1} \right\} \left\{ (\Delta p/bar)/(s/mm) \right\} \quad (8.3)$$

where P is the permeation coefficient, in square centimeters per bar second; Δp is the difference in oxygen pressure, which is equal to the partial pressure in the ambient air of 0.2 bar; and s is the thickness of the coating in millimeters.

The permeation coefficient of most coatings can be established. The challenges arise in determining the criteria for judging the effectiveness of a new coating type or of a coating system for an extreme service condition, where temperature, the electrical voltage, or both may be factors. The coating resistance (r_u^0) of the coating system should be considered to avoid blistering, and the formation of alkalis at cathodic polarization sites must also be considered (Table 8.1). Selecting the most suitable system to meet the project's specific demands should involve an evaluation of coating stability. The stability of a coating in these conditions must be tested in the laboratory to determine the suitability of the specific coating system in the given environment.

Stability refers to the failure of a coating system as a result of poor resistance leading to blistering, cathodic and anodic blistering, or disbonding. This chapter further discusses defects and failures at a later point.

Organic coatings include paints, plastics, and bitumen.

As stated earlier, these coatings can be applied as a thin layer of coating. These coatings are also applied as a base for a thick-layer coating system. Thin organic coatings are applied as liquids or as powdered resin and may be about 0.5 mm thick. In present day coating applications, this group is represented by epoxy resin coating, which is often abbreviated as EP coating. Such thin coatings contain many polar groups that promote adhesion.

The thick coatings are those that are over 1 mm thick and come from the polyolefin and bitumen groups. Polyethylene (PE) and polyurethane coatings are polyolefin coatings and tar coatings are examples of bitumen

Table 8.1 Typical Coating Resistance

Coating Type Unit →	Specific Resistance of Coating ρ_D^a Ω cm	Thickness of Coating s mm	Calculated Resistance Value Based on the ρ_D r_u^x Ω cm²	Resistance Value Obtained from the Laboratory Testing r_u^{ob} Ω cm²	Thickness of Coating s mm	Resistance Value Obtained from the Structure in Service Media r_u^c Ω cm²
Bitumen coating	$>14^{10}$	4	4×10^9	3×10^5	4 to 10	-10^4
Polyethylene	14^{18}	2	4×10^{13}	10^{11}	2 to 4	-10^5
Epoxy Resin coating	14^{15}	0.4	4×10^9	10^8	0.4	-10^4
Polyurethane coating	3×14^{14}	2	6×10^9	10^9	2.5	

r_u = Change in voltage/Change in current $\times s$ ($\Delta V/\Delta I \star s$)

coatings. The low-polar or no-polar thick coating systems are often combined with polar adhesives to achieve the required bond to the substrate.

Often a combination of polyolefin and bitumen coatings are used as well. For example, EP tar, or polyurethane tar may be employed. Similarly, epoxy-based polyethylene or polyurethane coatings are also used as high-efficiency coating systems.

Organic coatings have following properties.

i. Organic coatings have high mechanical resistance and high adhesion. These properties allow for the safe transportation and handling of coated metals. Mechanical damage to the coating is still a possibility, and it should be prevented. So, if damage to coating occurs, it should be repaired. If a damaged spot is left unrepaired and the steel substrate is exposed to electrochemical reactions, the steel substrate will corrode. Reference Figure 8.1 for an illustration of a holiday, or coating failure.

ii. Organic coatings have better chemical stability, allowing the coating to function effectively as the applied system ages.

iii. Organic coatings work very effectively with electrochemical corrosion protection systems. Often a combination of a coating with a cathodic protection system is designed to protect a high-value and high-risk structure from corrosion.

iv. Organic coatings have low permeability for corrosive components. However, each coating and the corrosive environment have to be carefully evaluated for the project-specific levels of tolerance. The permeability of a coating allows molecules of corrosive elements, such as O_2, H_2O, and CO_2, to migrate through the coating to reach the substrate. It also facilitates the migration of anions and cations to the surface of the steel. The driving force for the transport of all particles is a change in the electrochemical potential, which is related to the partial molar free enthalpy and electric potential. The electrical voltage involved in the process ranges from a few tenths of volts to several volts and arises from the polarization reactions, which are briefly discussed.

If *anodic polarization* occurs, exiting stray current or contact with foreign cathodic structures could be the cause.

If *cathodic polarization* occurs, entering stray current or cathodic protection could be the cause.

In anodic polarization, ion migration then causes electrolytic blister formation, which manifests on the coated surface (steel substrate) as pitting.

Cathodic polarization manifests as no change or a formation of oxide layer with annealing color.

The ion migration also creates a cell where the coated substrate acts as a cathode and the exposed metal (holiday in coating) acts as an anode. (For more information on reactions at the coating holiday, refer to Figure 8.1). At the metal and coating interface, the cathodic partial reaction of oxygen reduction as shown in Equation (8.4) is much less restricted than the anodic partial reaction shown in Equation (8.5)

$$O_2 + 2H_2O + 4e^- = 4OH^- \qquad (8.4)$$

$$Me = Me^{Z+} + Ze^- \qquad (8.5)$$

In the free corrosion activity at the rim of the holiday and the cathodic protection of the entire substrate in an electrolyte, the oxygen reduction and production of OH^- ions take place on the entire exposed surface. This reaction is similar to the one described in Equation (8.4). The pH of the electrolyte is significantly increased. The OH^- ions are in a position to react with adhesive groups in the coating, and this allows the migration under the coating. This migration results in *cathodic disbonding*.

The coating resistance (r_u^x) and relation to the required protection current is another aspect of coating efficiency in corrosion protection. Coating-specific resistance (ρ_D) is an important factor, along with the surface area (S), and the thickness of coating is indicated by letter (s) in the calculation of coating resistance r_u^x, which is obtained by calculation according to the following relationship.

$$\{r_u^x/\Omega m^2\} = 10^{-5}\{\rho_D/\Omega cm\}\{s/mm\}\dots \qquad (8.6)$$

Table 8.1 gives a comparative description of specific coating resistance.

The coating resistance value of a defect-free coating in an electrolyte is r_u^0, and coating resistance r_u is the value on the surface where the coating has defects (holidays). The current and potential measurements are used to derive the r_u values, as per following equation:

Resistance of coating with defect (r_u)
= (change in Voltage/change in current) * coating surface area
Thus,

$$r_u = (\Delta V/\Delta I) * S \qquad (8.7)$$

For high-efficiency coating system, the value of r_u should be very small, making these coatings perform well in demanding services. This is only possible with proper selection of the coating system and a very high quality application process that adheres to the established application parameters and inspection and test protocols. The resistance of a coating with holidays (r_u) is considerably lower than the established coating resistance value (r_u^X) or the coating resistance value of a defect-free coating (r_u^0) that is in an electrolyte. Needless to say, the lower resistence results from the in-service coating having developed some pores and defects. The resistance of the defect(s) is obtained from the sum-of-pores resistance R_F, where the coating thickness is s, diameter of the pore is d, and the medium's specific resistance is ρ, as shown in the below equation.

$$R_F = \rho\left(4s/\pi d^2\right) \tag{8.8}$$

The coating resistance value (r_u^X) is one of the most important pieces of data to obtain about the performance of high-efficiency coating. The coating resistance (r_u^X) is affected by the water absorption and increasing temperature that affect the resins used in the coating application.

The cell formation due to pores can be detected by measuring the potential of the coated surface. Defective coating will show more negative potential than a defect-free coated surface. The r_u^0 value of a thin coating system can range from 10^2 to 10^7 Ω m^2, and these values highlight the importance of the type of prime-coating and coating thickness. The ratio of resistance for a coating with holidays (r_u) and the coating resistance value (r_u^0) also have a great effect on a coating's performance. Thus, sufficient external protection for installations, even if cathodic protection is provided, is the best way to extend the life of structure and avoid failures.

In thin coating systems, blistering of the coating around defects is often observed. The conductivity of ions of the coating material causes the blistering. As is evident, blistering involves an electrochemical reaction, and hence, it is also called electrochemical blistering. When exposed to electrolytes for a significant period of time, thin coatings that have low resistance (r_u^0) values often show blistering flaws. This becomes especially true if the electrolyte is seawater or a similar saline solution. In such environments, alkaline ions are present, and the permeation of water and oxygen is possible, resulting in the formation of OH^- ions at the interface of the metal and coating due to a cathodic partial reaction. The reaction produces a caustic solution with migrating alkaline ions, allowing water molecules to diffuse at the reaction sites through osmosis and electroosmosis, thus causing the coating to blister.

ANODIC BLISTERS

Anodic blistering forms when a foreign cathodic object is in contact with the coated surface within the electrolyte. In this situation, the anions form soluble corrosion products with the cations of the substrate metal—the cations are formed due to the polarization of anode—and water molecules are prevented from migrating due to osmotic and elctroosmotic processes. This results in relatively smaller blisters, which is a typical attribute of anodic blistering. These anodic blisters may conceal pitting of the substrate, and anodic blisters are easy to miss during an inspection as compared to large size cathodic blisters.

Blister distribution and formation are connected to ion conductivity in the coating material. One of the important aspects of the high-efficiency coating application process is proper surface cleaning and preparation prior to the application of the coating. The cleaning includes the removal of salt residue from the substrate. This is a very important step because the presence of salt residue enhances the electrochemical blistering process, especially for primer coatings that contain ions.

In anodic blistering, the disbonding is strongly suppressed, as the partial reaction dominates at the defect, leading to the formation of OH^+ ions in anodic polarization. The metal loss causes pitting rather than uniform corrosion under the coating.

CATHODIC BLISTERING AND DISBONDING

Alkaline silicate primers are especially prone to cathodic blistering. Mechanical damage on the pipe or structure substrate can also cause blistering of the coating. Blister attack increases with increasing cathodic polarization in the damaged area, and this leads to the cathodic disbonding of the coating. Disbonding can also occur with long-term cathodic polarization in solutions without alkaline ions or in coatings with larger low resistance (r_u^0) values, where the reaction of the blister liquid is neutral. In large-area adhesion loss, the presence of humidity is only indicated by slight rusting, and the area around the anode is particularly noted for its dryness. The electrolysis process is not involved in this type of disbonding. The process involved is referred as Cohen's rule, which deals with changes in a negatively charged macrogel coating material in relation to water.

Cathodic disbonding is caused by the production of OH^- ions in pores or damaged areas of the coating. The high concentration of ions is possible only if counter ions, such as NH_4^+ and Ba^{2+}, are also present. Some very limited cases of disbonding are also reported in the presence of Ca^{2+} ions. The depth of disbonding increases with the increasing concentration of alkaline ions. Disbonding even occurs in a potassium chromate (K_2CrO_4) solution. The disbonding rate decreases over the time as the OH^- ions are consumed by reactions with adhesive groups in the coating system. Temperature increase and larger holidays also increase the disbonding process. The type of coating is another factor that affects the disbonding, and many polar adhesives groups, such as resin coatings, fare better over thick layer coatings. This performance of resin coatings contrasts with the coating behavior in blistering conditions discussed earlier. The application of polar adhesives combined with thick coating systems, such as three-layer polyethylene or multilayers of polyurethane coatings, are considered to be high-efficiency coating systems.

The coating blister population and potential curves can intersect, making the short term assessment of the effects of cathodic over protection difficult to measure. The possibility of blister formation generally increases with over protection, if the potential is less negative than -0.83 V. Both cathodic and anodic blisters can coexist in same general area in a free general corrosion.

ENAMEL COATING

Enamel coating protects the substrate as a barrier from the corrosive environment. This coating is hard and can chip if hit hard, making it very easy to damage, especially during transportation and handling. If the enamel coating is damaged, the corrosion protection can be lost. The coating is applied in several layers, including a primer that serves as an adhesive supporting the top coats that are essentially for corrosion protection.

Enamels come with various chemical compositions, resulting in varied chemical stability. Each enamel system must be tested and evaluated for the specific environment that it is supposed to protect against.

The reaction between the interface of steel and enamel is very sensitive to hydrogen recombination, however, leading to cracks and spalling. The hydrogen can come from the enamel itself, or it can be the byproduct of the process. Tests must be conducted before selecting any specific type of enamel coating for the given service condition.

If the correct type of enamel is not selected, the damage to the top layers of the coating can expose the adhesive layer to cathodic polarization alkalis, and this exposure has ability to enlarge the effective size of a defect. This type of coating is not well-suited for salt-rich media, and the negative effects of the media on the coating can be pronounced if small pores are present in the coating.

Despite such drawbacks, enamel coatings are electrochemically inert. When not perforated by the process or chipped during handling, enamel is impervious to water. But the coating's minute pores can be difficult to identify. Holiday testing often reveals small pores covered with conducting oxides, which are the products of corrosion; this essentially stops further corrosion. If the area of a holiday is larger, the corrosion protection offered by the production of conducting oxides is lost.

Steel structures, such as tanks, bulk heads, and fittings, that show cathodic affectivity are often protected by enamel coatings. The holidays in the enamel coating present the possibility of pitting corrosion. Cathodic protection provides additional protection at holiday locations.

METALLIC COATINGS

Metallic coatings are applied in cases where the substrate is coated with a more noble metal, as with copper on steel. This type of protective coating is effective only when the coating is free from pores or damages. The defects in a metallic coating can cause severe damage to substrate via the formation of local cells.

Another variant of metallic coating is the coating of a substrate with a less noble material, as when steel is coated with zinc, or when aluminum is thermally sprayed on steel risers or offshore fixed structures. This type of coating is effective only when the corrosion product of the metal coating is able to restrict the corrosion process, and formation of aeration cells is stopped by the coating. The possibility of damage due to blistering and cathodic corrosion exists, and it must be addressed during the design stage, before selecting this type of metallic coating. The absorption and recombination of hydrogen is possible, and that can lead to blistering and cathodic corrosion. For both zinc and aluminum, additional protection can be provided by cathodic protection. The polarization properties and low protection current should be determined per each specific case. However, the protection potential must be limited toward the negative end. This limiting of the protection potential can, in turn, limit the formation of aluminates and zincates as corrosion products.

THERMAL SPRAY ALUMINUM COATINGS

Thermal spray aluminum (TSA) coatings have been used successfully to protect the exterior of offshore structures and other process vessels, tanks, and equipment. The main application for these coatings has been seen in the treatment of steel risers in offshore oil and gas production systems.

Yet, TSA coatings have been used across the industry in various other applications, including in processing plants and other facilities in the oil and gas, petrochemical/refining, and paper industries.

There are a number of different ways to apply sprayed aluminum coatings using an oxy-fuel spray system or a plasma-arc spray system. The general processes has a number of variants, such as heat source use and wire applications method.

One of the application methods is the wire-arc system. This process uses a spool of aluminum wire that is heated within the spray gun by the oxy-fuel, and the molten aluminum is then atomized as it leaves through the nozzle under the compressed air pressure. This atomized liquid aluminum then deposits over the prepared surface of the steel substrate.

In the plasma-arc spray system, on the other hand, the feed of wire can be moved either by a pull or push mechanism, and as the wire passes through the spray gun, it is electrically charged, arced, and melted. Located immediately behind this hot plasma zone is a compressed air jet that blows across the arc area, atomizing and propelling the molten metal particles forward to coat the substrate.

The quality of the consumable, arc, and melting process and the preparation of substrate are the major factors that determine the quality of the system. For aluminum wire, the purity is maintained to above 99.5%.

Compressed air is kept clean and free from oil and moisture. Blast cleaning abrasives must also be pure enough to obtain a surface profile of Sa3, and the operator must monitor ambient conditions check, environmental cleanliness, dew point, and relative humidity.

A typical display of plasma wire spray system is shown in Figure 8.2.

Often applied TSA coating can be as thick as 300 μm (or 12 mils), and this thickness gives the coatings the following advantage over conventional coating systems.

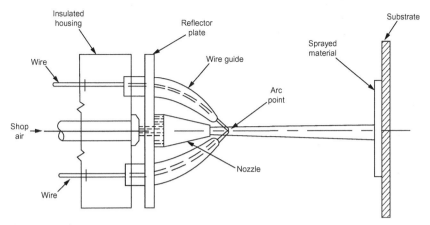

Figure 8.2 Plasma arc spray system.

- Excellent adhesive and cohesive strength of coating
- No requirement for cathodic protection
- Corrosion protection
- Lifetime of service (Thicker coatings provide longer life, based on the sacrificial wear rate of the metal)
- Harder, better adhesion for high temperature applications
- No mixing setup or pot-life limitations found with other coating systems

CHAPTER NINE

High-Efficiency Coating and Cathodic Protection

INTRODUCTION

Offshore structures are installed in an electrolyte (sea water, marine sediments, mud, etc.), or they are exposed to the splash zone. As a result, these structures are subjected to severe corrosion attacks. Offshore installations require a most effective corrosion control system, which may include coating, lining, cathodic protection, or any combination of these measures. The coating system is developed in conjunction with the applicable cathodic protection system. The design of the CP system must also be based on the performance of the coating system selected and applied. Both external and internal corrosion protection of pipelines and risers should be considered.

In general, pipelines and structures are coated with high-efficiency coating systems composed of fusion-bonded epoxy, two layers of extruded polyethylene, and three heavy layers or multilayers of (polyolefin) polyethylene or polyurethane coatings. These systems have performed very well, but their success depends on correct application and quality control in terms of surface preparations, surface temperature, coating application procedure, etc.

Tables 9.1 and 9.2 describe the typical application of a seven-layer coating to a flow line and risers in a deep water system.

The coating industry has continuously strived to create higher quality products that meet the demands posed by challenging environments, such as colder temperatures and deeper sea levels. New coating systems that claim meet these challenges are appearing and are often given more attractive names based on their specific properties.

In some cases, improvements on existing coating systems are achieved by eliminating the sharp well-defined interface between the FBE and adhesive layers or between the adhesive and outer polyethylene or polyurethane layers. The removal of interfaces and intermingling of the layers disperse the defining layers, thus effectively making the system a single-layer coating

Table 9.1 Typical High Efficiency Coating Applied to Flow Lines

Layer	Coating Layers	Coating Thickness
1	EPOXY (FBE)	300 μm
2	Adhesive	300 μm
3	Solid polypropylene	8.4 mm
4	Solid polypropylene	30 to 40 mm
5	Polypropylene shield	4.0 to 5.0 mm
6	Polypropylene foam (TDF)	30 mm
7	Polypropylene outer shield	5.0 mm

Table 9.2 Typical High Efficiency Coating Systems Applied to SCR (Risers)

Layer	Coating Layers	Applied Coating Thickness
1	EPOXY (FBE)	300 μm
2	Adhesive	300 μm
3	Solid polypropylene	8.4 mm
4	Syntactic polypropylene (SPP)	23 mm
5	Polypropylene shield	5.0 mm
6	Polypropylene foam (TDF)	23 mm
7	Polypropylene outer shield	5.0 mm

system. As these types of coatings are powder-applied, the coating is able to avoid the quality issues generally associated with such systems, including uneven application on welds and the possibility of air pockets at the edges of the weld profiles, often referred as tenting.

For the pipelines and for riser sections in the submerged zone, external corrosion protection is normally achieved by a thick film coating in combination with cathodic protection. In selecting the most cost-effective strategy for corrosion control, designers must use life cycle cost analysis to assess all major costs associated with the operation of the pipeline system, as well the investment costs for corrosion control. When fluid corrosivity and the efficiency of corrosion mitigation cannot be assessed with a high degree of accuracy, a risk cost may be added for the specific option being evaluated. The risk cost is the product of the estimated probability and consequences of a particular failure mode. The probability of such failures should reflect the designer's confidence in estimating the fluid corrosivity and the efficiency of the corrosion control options being evaluated. Depending on the failure mode, the consequences of failure may include costs associated with increased maintenance, repairs, and lost capacity, as well as secondary damage to life, environment, and other investments.

As stated above, coating is one of the most important methods for preventing corrosion and protecting investment, and in the previous chapter, we discussed the basic principles associated with various general types of coating. We also learned that no coating, however efficient, is capable of preventing corrosion of the substrate in all environments. While a coating system must have certain specific properties, not all coatings are suitable for all environments. Coating properties are also discussed in the previous chapter.

The selection of a coating should consider the properties of the coating and the following design-specific factors.

- Type of environment
- Operating temperature
- Ambient temperature and humidity at the time of application
- Accessibility of application
- Geographical and physical location of service
- Surface preparation methods that can be applied and resources available
- Handling and storage
- Total cost of coating

A purpose-specific coating specification should address the above-listed factors in the utmost detail. Table 9.3 gives some the common coating systems used. Wherever possible, designers should also refer to the applicable industry specifications for guidance and compliance. A well-prepared specification addressing the environmental factors and project purpose must be provided to the coating contractor, and each point with the specifications must be discussed and agreed upon, including the basis for the selection of a specific coating system, as well as the requirements for the testing and inspections to be carried out on the coating system during and after the application process.

Several international bodies have developed specific documents to address the coating and cathodic protection of offshore structures, and local regulatory bodies have mandated compliance with some of these guidelines, while others are used as recommended practices. These guidelines are listed in Section 5 of the book.

The external coating for the corrosion protection of linepipe is a factory-applied coating system consisting of multiple layers. These coatings may also have thermal insulation on them. Some coating systems may further include an outer abrasion resistant layer for mechanical protection, primarily during laying in rocky trenches, rock-dumping, or pulling through a bored hole, as might occur with a directional drilling through the crossing of waterways, major highways, or railroads.

Table 9.3 Some of the Common Coating Systems Used

Coating Type	Properties	Limitations
Coal tar enamel	Minimum holiday, low current requirement, good adhesion to steel substrate, good resistance to cathodic disbonding,	Health and air quality hazard, change in allowable reinforcements, limited applicators.
Mill –applied tape coatings	Minimum holiday, Ease of application, low energy required for application, good adhesion to steel substrate.	Shipping and installation restrictions, UV and thermal blistering, shielding CP from soil, stress disbonding.
Crosshead-Extruded Polyolefin with asphalt or butyl adhesive	Minimum holiday, Low current required, Ease of application, Low energy required for application, Non-polluting.	Minimum adhesion to steel, Limited storage, Tendency to tear along the length of the pipe.
Dual side extruded Polyolefin with butyl adhesive	Minimum holiday, Good adhesion to steel, Ease of application, Low energy required for application, Non-polluting, Excellent resistance to cathodic disbonding.	Limited application,
Fusion Bonded	Low current required, Excellent resistance to cathodic disbonding, Good adhesion to steel, Resistant to hydrocarbons.	Higher application temperatures, steel and coating interface imperfections, Higher moisture absorption, Lower impact and abrasion resistance.
Multi-layered epoxy/ extruded polyolefin systems	Lowest current required, Excellent adhesion to steel, High impact and abrasion resistance, High resistance to cathodic disbonding, Excellent resistance to hydrocarbons.	Limited application, Exacting application parameters, Higher initial cost, Possible shielding of CP current.

The field-joint coatings (FJCs) are the coating system that is applied to protect girth welds, and this term is used irrespective of whether the coating is applied in the field or on a spool base on pipelines and risers for reel laying. These coatings may be single-layer or multilayer coating systems.

> ## PIPELINE EXTERNAL COATINGS

The external linepipe coating system is selected based on a consideration of the following coating properties.

- Corrosion-protective properties:
  - **i.** The permeability to water
  - **ii.** Dissolved gases and salts
  - **iii.** Adhesion
  - **iv.** Freedom from pores
- Mechanical properties demanded by installation and operation
- Resistance to physical, chemical, and biological degradation, both in service and during storage and transportation
- Operating temperature range and design life
- Ability to meet with the rigors of fabrication and installation procedures
- Ease of FJC and field repairs
- Compatibility with concrete weight coating where concrete coating is applicable
- Compatibility with cathodic protection, given reducing current demand for cathodic protection
- Thermal insulation properties where required
- Environmental compatibility and health hazards during coating application, fabrication, installation, and operation stages

The functional requirements of the required coating system should be defined in the specification. Project-specific requirements regarding quality control will be described. The following are the general properties that may be specified.

- **a.** Maximum and minimum thickness
- **b.** Tensile properties
- **c.** Impact density
- **d.** Adhesion
- **e.** Flexibility
- **f.** Resistance
- **g.** Cathodic disbonding resistance (potential and time factored)
- **h.** Thermal resistance or conductivity
- **i.** Abrasion resistance
- **j.** Electrical resistance
- **k.** Cutbacks
- **l.** Resistance to hydrostatic pressure

COATING MATERIALS, SURFACE PREPARATION, AND APPLICATION

The coating manufacturer should be asked for documented proof that the system possesses the required coating properties. A coating manufacturing qualification should be executed. Where possible, the testing should be observed to ensure that specified testing protocols are followed, and the results should be reviewed for specification compatibility and accepted before the coating work starts. Testing and property review are especially important for novel products, given that the relative lack or ambiguity of manufacturing and field data associated with coating performance.

If different from the manufacturer, the coating applicator must qualify a manufacturing procedure specification (MPS), which should agree with the manufacturer's recommendations. Once approved, all coating work should follow the stated procedure. The MPS should include evaluation of following key stages and parameters.

a. Coating materials
b. Surface preparation
c. Coating application
d. Inspection and testing
e. Coating repairs
f. Handling and storage of coating material during and after application

Upon approval of the MPS, the applicator should submit a quality plan and an inspection and test plan (ITP) for the review, prior to comments and final acceptance. The quality plan and ITP should define details of application and preparation methods, frequency of inspection, acceptance criteria for testing, and calibrations. The document must reference the applicable specifications and procedures for inspection, testing, and calibrations. The document should also describe in clear terms the handling of nonconforming coating materials and products. In addition, the ITP should identify the deliverables at each stage of the inspection. This set of documents is intended to establish the guidelines for the performance, testing, and acceptance of the applied coating.

Riser sections that are installed in the atmospheric zone are protected from external corrosion by a proper coating with surface preparation and coating application according to the applicable standard.

A riser installed in the splash zone requires a thick film of coating. The steel risers installed above the lowest astronomic tide (LAT) may be given an additional corrosion allowance, and, as the cathodic protection may not be functioning efficiently, these risers may also be provided with external cladding or lining with corrosion-resistant alloys (CRAs), such as cupronickel alloy seething.

To prevent corrosion of the risers contained in J-tubes, conductors, tunnels, and equivalent structures, the annulus should be filled with noncorrosive fluid and sealed at both ends. Provisions for monitoring annulus fluid corrosivity should be considered. Internal corrosion protection, and especially temporary protection techniques, should be used to safeguard the internal surface of the pipe when the annulus is filled with fluids. The properties of the fluid and its possible reaction to the steel should be considered and evaluated.

INTERNAL CORROSION PROTECTION

Internal corrosion control, including the temporary corrosion control, is briefly discussed below.

The portions of fixed offshore structures that get flooded during installation or to stabilize the structure may be treated with oxygen scavengers and capped to prevent any further environmental reactions that might start corrosion.

Most fluids transported in pipeline systems are potentially corrosive to ordinary linepipe steel. As a result of this corrosiveness, the selection of a coating system for internal corrosion protection of pipelines and risers has a major effect on the detailed design of those structures and should be part of the conceptual design evaluation process. Depending on the corrosion severity, the following options for corrosion control can be considered. One or a combination of these measures may be used to protect the internal surfaces of pipelines, risers, bulk heads, and other equipment and storage facilities included in offshore structures. Some facilities that require long-term protection and it is practically possible to install a CP system they may even be given a cathodic protection system with anodes placed inside.

- Processing of fluid for the removal of liquid water and/or corrosive agents
- Use of linepipe with internal metallic lining or cladding with intrinsic corrosion resistance
- Use of organic corrosion-protective coatings or linings often in combination with fluid processing or chemical treatment

- Chemical treatment, including the addition of chemicals with corrosion-mitigating functions
- Addition of a suitable corrosion allowance

TEMPORARY PROTECTION FROM CORROSION

For temporary protection from internal corrosion during storage, transportation, and flooding, the techniques include end caps, rust-protective oil, and wax. For sections that are flooded for some time, chemical treatments, including the use of biocide, oxygen scavengers, or both, are used. The use of a biocide to treat flood water is very important, even if the flooding is for a short duration, as the incipient bacterial growth established during flooding may proceed during operation and cause corrosion damage.

For uncoated steel pipelines, an oxygen scavenger is not used, because oxygen will get dissolved in seawater, and the remaining oxygen on the steel surface will be rapidly consumed by uniform corrosion. This uniform corrosion, however, will not cause any notable loss of wall thickness. Film-forming or passivation corrosion inhibitors are not actually required, and the harmful effects of some inhibitors on the steel substrate should always be evaluated before their use.

COATINGS FOR RISERS

As described above, corrosion conditions differ in different zones, and these differing corrosion conditions demand specific attention regarding the selection of coating type and application.

In the submerged zone and in the splash zone below the LAT, an adequately designed cathodic protection system provides an additional protection measure. Added cathodic protection is capable of preventing corrosion in any areas of damaged coating on the riser. In the tidal zone, cathodic protection will be only marginally effective, though.

Adverse corrosive conditions also prevail in the zone above LAT where the riser is intermittently wetted by waves, tide, and sea spray (splash zone). Particularly severe corrosive conditions apply to risers heated by an internal fluid. In the splash zone, the riser coating may also be exposed to mechanical damage by surface vessels and marine operations, while there is limited accessibility for inspection and maintenance.

The riser section in the atmospheric zone above the splash zone must be protected by coating as well. This zone is relatively shielded from both severe weathering and mechanical damage, however, and it is more accessible for inspection and maintenance activities.

WHAT IS A SPLASH ZONE?

A splash zone is defined as follows:

The splash zone refers to the external surfaces of a structure or pipeline that are periodically in and out of the water due to the influence of waves and tides. The height of a splash zone is the vertical distance between splash zone upper limit and splash zone lower limit.

The wave height of a splash zone is based on the probability of being exceeded equal to 10^{-2}, as determined from the long-term distribution of individual waves. If this value is not available, an approximate value of the splash zone height may be represented by significant wave height with a 100-year return period.

For a specific riser, the division into corrosion protection zones is dependent on the particular riser or platform design and the prevailing environmental conditions. The upper and lower limits of the splash zone should be determined to ensure correct corrosion protection design. Generally, these data can be obtained from the well-maintained marine records and historical records of the area.

Different coating systems may be applied in the three corrosion protection zones defined above, provided those systems are compatible with the environment and the adjoining coatings. The protective properties of the coatings for all three zones must meet the requirements introduced in this chapter's discussion of coating properties. The properties of fastening devices for risers should also be selected on the basis of their compatibility with the riser coating.

The following additional properties of coatings used in splash and atmospheric zones may also be considered.

- Resistance to under-rusting at coating defects
- Maintainability
- Compatibility with inspection procedures for internal and/or external corrosion
- Compatibility with equipment
- Procedures for the removal of biofouling, where the potential for biofouling is identified
- Fire protection, if a fire hazard is identified

The use of a corrosion allowance to compensate for external corrosion due to coating damage can also be considered for the protection of the splash zone. The need for, and benefits of, a corrosion allowance depends on the type of coating, corrosive conditions, design life, consequences of damage, and accessibility for inspection and maintenance.

In selecting a coating for the submerged zone, the designer must still ensure that the coating system demonstrates the discussed properties. In addition, resistance to biofouling is relevant in the portion of the submerged zone that contacts surface water, as well as in the lowermost section of the splash zone. Mechanical and physical coating properties are also relevant for riser coatings, dependent on the particular corrosion protection zone. The applicable requirements to properties for each coating system and for quality control shall be defined in a purchase specification.

External cladding with certain Cu-Ni-based alloys is used for both corrosion protection and antifouling, primarily in the transition of the splash zone and the submerged zone. Certain CRAs may not need any such coating. However, metallic materials with antifouling properties must be electrically insulated from the cathodic protection system to be effective. Multilayer coatings and thermally sprayed aluminum (TSA) coatings are also used to protect the atmospheric and submerged zones, and in the splash zone as well, if functional requirements and local conditions permit.

COATING MATERIALS, SURFACE PREPARATION, AND APPLICATION

Riser coatings may be applied after fabrication welding, and in the atmospheric zone, after installation. If the exact location is predetermined, then such coating may also be applied prior to the installation.

All coating work is carried out according to a qualified procedure. The coating MPS details the requirements for handling, storage, identification marking, and inspection of coating materials.

Chapter 10 in this section details the requirements for the qualification of the coating manufacturing process, coating MPS, and quality plan. The full spectrum of coating qualifications involves the qualification of product quality, the assessment of the coating application procedure, and the testing of the applied coating.

SURFACE PREPARATION FOR THE COATING

The importance of surface preparation for coating cannot be over emphasized. The objective of surface preparation is to provide the necessary degree of cleanliness and the specific anchor pattern required for the designed coating system. Automated blast cleaning is the most desired method of surface preparation. The abrasive material used for cleaning should have specified material size and hardness in order to produce the surface profile required for the chosen coating system. Surface profile tests are periodically conducted to ensure that the surface is ready to receive the coating.

When an air compressor is used for blasting, especially in field application or maintenance coating, the air supply should be able to maintain a minimum of 100 psi pressure to the nozzle, and the equipment should be able to separate oil and moisture from the air supply line.

Surface preparation level is specified for the specific coating system. Several regulatory and voluntary bodies, such as NACE, SSPC, International Organization for Standardization (ISO), and Swedish industry authorities, have produced elaborate descriptions of surface preparation levels that should be consulted and specified with the selected coating system. The data sheet for a specific coating system details the requirements for surface finish, surface profile, salt contamination level, and steel temperature prior to surface preparation and coating application.

The substrate temperature for the application of a coating is an important variable that is specified for each coating type. The air temperature for the application of a coating is also an important variable, and often this temperature is set at a minimum of 5°F above the dew point.

PROTECTION OF SPLASH ZONE OF OFFSHORE FIXED STRUCTURES

Splash zone corrosion control for offshore fixed structures is safeguarded by cathodic protection measures supported by the application of high-performance coating. Additional wall thickness and sheathing of CRAs are also used. There are several coating system options available, and the project design requirements should guide the selection of a suitable coating system.

HIGH-BUILD ORGANIC COATINGS

High-build organic coatings consist of silica-glass flakes or fiberglass, and these coatings are applied to a thickness of about 40–120 mils (1–5 mm), with an overcoat applied for antifouling.

HEAT-SHRINK SLEEVES

Heat-shrink sleeves are plastic sleeves that are internally layered with sealant adhesives. The surface preparation removes the rust and coarse roughness of the surface. These sleeves are then wrapped around the substrate with wrap overlap onto the straight section of the structural member, thus covering the surface to be protected. The plastic shrinks with the application of higher heat, sealing any gap that is within the wrapped area. This seal is held tight by the shrink force of the sleeve, and sealant spreads to the voids, if any exist. These types of coatings have performed with some limited success in different environments across the world.

HIGH-PERFORMANCE COATINGS

High-performance coating systems used in the atmospheric zones also protect the splash zones with some success, and in most cases, they are used in combination with applied cathodic protection. A coating thickness of 10-20 mils (250-500 μm) is often applied.

The success of high-performance coating systems in the splash zone of long-life structures has produced mixed reports, however.

THERMALLY SPRAYED ALUMINUM COATING

TSA is a coating that uses an aluminum matrix. It is produced by flame or arc-spraying a solid wire that is heated, atomized, and sprayed.

The procedure for applying TSA for the corrosion protection of steel includes following key steps:

(a) Proper surface preparation of the substrate steel

(b) Proper application of the TSA

(c) Proper application of the sealer or sealer and topcoat where specified. The surface preparation procedure includes the use of abrasive blasting, thermal spraying, sealing or top-coating, and in-process quality control. The TSA coating process is discussed in more detail in Chapter 8.

Surface Finish

The steel substrate is prepared to white metal finish, meeting NACE No. 1 or SSPC-SP 5 specifications for marine and immersion service, or to near-white metal finish, meeting NACE No. 2/SSPC-SP 10 specifications for other service applications. The level of soluble-salt contamination on the surface is controlled through the evaluation of salt possibility and included in the contract. Visual inspection of the surface is also conducted to ensure that the surface finish and cleanliness meets the requirements of SSPC-VIS 1 standards.

The objective of surface preparation is to establish the angular profile depth on a steel substrate to a minimum depth of ≥ 65 µm (2.5 mils), with a sharp angular shape. This measurement is taken periodically, according to the NACE Standard RP0287 or ASTM D 4417 Method C, which uses a replica tape, x-coarse, 38-113 µm (1.5-4.5 mils), or Method B, which uses a profile depth gauge. Often both methods are used to verify and supplement the measurement process.

When surface preparation is carried out by blast cleaning, the *Manual Blasting* method profile measurement is conducted, and at a minimum, one profile depth measurement is taken for every 1-2 m^2 (10-20 ft.^{2}) of blasted surface.

For an *automated blasting system*, one profile depth measurement is carried out per 100-200 m^2 (1000-2000 ft.^{2}) of blast surface.

The media used to obtain the angular blast profile are clean, dry mineral and slag abrasives that are evaluated per SSPC-AB 1, recycled ferrous metallic abrasives evaluated per SSPC-AB 2, and steel grit abrasive evaluated per SSPC-AB 3. The media is kept free from oil contamination. The soluble salt contamination is measured according to the procedure given in ASTM D 4940, and the suitability of the angular blast media, blasting equipment, and blasting procedures is validated.

The TSA tensile bond is measured according to the procedure given in the ASTM D 4541, using a self-aligning adhesion tester. The minimum tensile bond value is given in Table 9.4, and the table also includes other types of thermally applied coating.

The bend test is a macro-system test of surface preparation, equipment setup, spray parameters, and application procedures. As a quality control test, a bend test is carried out to determine the bond. The test involves a 180°

Table 9.4 Minimum Tensile Bond of TSA Coating Systems

Feedstock	MPa(psi)
Zn	3.45(500)
Al	6.89(1000)
85/15 Zn/AL	4.83(700)
90/10 Al$_2$O$_3$ MMC	6.89(1000)

bend on a mandrel diameter based on the TSA thickness. This is a qualitative test of the ductility and tensile bond of the TSA.

TSA is used for the protection of the splash zone, and the application is conducted in specialized shops using either a flame spray process or an arc-spray method. The molten aluminum is atomized through a high-pressure-control spray-nozzle and deposited in a uniform layer of about 8 mils (200 μm) on the substrate. The adhesion of aluminum to the steel substrate is of vital importance, and quality control tests require adhesion at a minimum of 1000 psi (7000 kPa), tested in accordance with ASTM C633. The applied aluminum may contain some pores, and is prone to further damage. So, to protect the coating, a spray of silicon-based sealant is applied to the coated surface.

FIELD-JOINT COATINGS

When selecting a FJC or a repair coating system, care should be taken to select a system that is compatible to the mill-applied coating system, and as far as possible, external coatings must match the properties of the mill-applied linepipe coating system. Where compatibility is not established, the design of the CP should be able to step in and compensate for the deficiency. The risks associated with hydrogen-induced cracking by cathodic protection should be kept in mind and accounted for in the CP design.

For pipes with a weight coating or thermally insulated coating, the FJC is typically made up of an inner corrosion protective coating and an in-fill. The objective of the in-fill is to provide a smooth transition to the pipeline coating, as well as mechanical protection for the inner coating. For thermally insulated pipelines and risers, the in-fill should also have adequate insulating properties.

Selection of an FJC requires the considerations described earlier for pipeline and riser coatings. In addition, sufficient time for application and hardening or curing is crucial during barge-laying of pipelines.

Preferably, riser FJCs have properties that match the selected pipe coating, and in the splash zone, FJCs should be avoided unless it can be demonstrated that their corrosion protection properties are closely equivalent to those of the adjacent coating.

Relevant coating properties are to be defined in a project specification. If possible, the specification should require that the FJC have the same properties as the involved pipelines and risers.

COATING MATERIALS, SURFACE PREPARATION, AND APPLICATION

The coating applicator should have a proven ability to meet the coating manufacturer's recommended application procedure for producing a coating that meets the specified properties. A qualification program that addresses all testing and inspections, including destructive testing of coatings, is to be prepared and tests conducted accordingly, prior to start of work, and the results of the tests are to be recorded and reviewed to the satisfaction of the project engineers. For systems that are applied at sea, the qualification program must include installation at sea, with subsequent destructive testing.

Furthermore, all coating work is carried out according with a procedure that is qualified and approved. The points that are to be addressed in the FJP are listed below. These points are also to be described in detail in the FJP MPS:

- Coating and in-fill materials
- Surface preparation
- Coating application
- In-fill application where applicable
- Inspection
- Coating repairs
- In-fill repairs where applicable

SURFACE PREPARATION FOR FJC

The surface that is to be coated for the field coating is normally prepared by local blasting grinding or brushing to a minimum NACE cleaning level of 3, or Sa 2, according to ISO 8501-1. Some coatings may be applied by brushoff cleaning to NACE 4 or Sa1 level.

The coating manufacturing procedure, as described above, should include the preparation process. The visual examination and nondestructive examination of FJCs should also be included. Where any of these examinations are not practical, relevant parameters affecting coating quality should be closely monitored.

CONCRETE WEIGHT COATING

The concrete coating of pipelines is not corrosion control. The description of such coating is included in this chapter because concrete

coating provides mechanical stability to the pipe which is coated with corrosion coating below the concrete. The objective of concrete weight coating is to provide negative buoyancy to the pipeline and to provide
mechanical protection of the corrosion coating during installation and
throughout the pipeline's operational life.

Requirements for raw materials, such as cement, aggregates, water, addi-
tives, reinforcement, and coating properties, are clearly defined in the project specification. The following properties are expected of concrete
coating.

- Calculation and compensation provided to achieve the submerged
  weight required for negative buoyancy
- Thickness of the coating
- Density of concrete to be coated
- Compressive strength of the concrete
- Level of water absorption in specified concrete
- Concrete's impact resistance
- Bending resistance-flexibility
- Required cutback of coating

Minimum requirements for some of the above-listed properties are to be
determined during the engineering stage and included in the project specifications. Similar requirements for steel reinforcement are also prepared on
the lines described below.

Project-specific requirements for quality control, including pipe tracking
and documentation, are also clearly described in the purchase documentation.

CONCRETE MATERIALS AND COATING MANUFACTURING

Before starting coating production, the coating manufacturer
shall document that the materials, procedures, and equipment to be
used are capable of producing a coating with the specified properties. A
preproduction test should be performed for documentation of certain properties, such as impact resistance and flexibility (bending strength).

All concrete coating work must then be carried out according to a qualified MPS. The following items are to be included in the MPS.

- Coating materials
- Reinforcement design and installation
- Coating application and curing
- Inspection and testing

- Coating repairs
- Handling and storage of coated pipes

The concrete constituents and manufacturing method should be selected to provide the following basic requirements based on as-applied coating properties:

- Minimum thickness: generally 40 mm
- Minimum compressive strength: an average of 3 core specimens per pipe reading a minimum of 40 MPa (as per ASTM C 39)
- Maximum water absorption: 8% by volume by test of coated pipe according to agreed method
- Minimum density of concrete: 1900 kg/m^3 (ASTM C 642)

CHAPTER TEN

Testing of High-Efficiency Coating Materials and Their Efficiency

INTRODUCTION

When a high quality product is expected to perform efficiently in very challenging environments, that product should be produced using a higher standard of quality control. High-efficiency coating products must protect metal substrates from corrosive elements in and above the ocean. So, the quality of these products must be verified to the highest degree possible. The process of strict quality control begins with the development of the product at the research laboratories of coating manufacturers, and these development activities are often closely guarded secrets. Several formulations and combinations are produced, formulated, reformulated, and mixed to achieve the desired properties. The quality control practices followed during this development phase are very stringent, and it is not uncommon for a product to require years of effort and multiple rejections, many of which send the scientists back to the drawing board to start anew.

This discussion does not focus on quality control during the research and development phase, however. Instead, it addresses the quality control procedures used to determine the properties of product batches, when the product has already been released in the market. These tests are used to verify that the selected product, or more specifically the selected batch or batches, meet a project's stated requirements. To this end, the quality control tests compare batch properties against internationally acceptable test specifications, which also outline test protocols to follow in order to obtain verifiable results that can be analyzed and compared.

Quality control tests can be classified into two distinct groups. One set of tests are conducted on the product itself, usually as a validation tool. The other set of tests evaluates the applied product in order to validate that the stated quality of the applied coating is in fact achieved during actual use. Some overlap exists between the two groups of tests based on specific project requirements. In general, the distinction between the two test types is very clear, however. The remainder of this chapter is divided into two parts to discuss product and production testing for coatings.

Product testing validates the behavior of the coating product in various test conditions and for various properties of the product. Product testing is primarily carried out to verify the properties of the coating material. These properties include the behaviors of product at various temperatures, the product's moisture content, and how the product gels. The testing of the applied coating validates the properties after the coating product is in use on the substrate. The tests are intended to determine the physical properties of the applied coating, and they include measurements of coating flexibility, and resistance to impact-related damages.

PRODUCT TESTS
Thermal Analysis of Epoxy and Cured Epoxy Coating Films

Polymer systems such as epoxy change their properties significantly when heated. The changes are mainly related to altered thermal, chemical, and electrical resistance, as well as mechanical strength. In actual practice these changes do not happen abruptly at one set temperature but over a time, but for the purposes of testing and evaluation, a set temperature is used, and this is called the glass transition temperature (T_g). Polymers that are exposed to a temperature below the T_g exhibit much greater dimensional stability, higher physical strength, electrical insulation, and chemical resistance, and all these properties are desirable in a high-performance coating system. These properties are significantly diminished at temperatures above T_g, however. So, the T_g is taken as the maximum sustainable operating temperature for a polymer system, such as an epoxy coating system. There are several ways to measure and determine the T_g of an epoxy coating system, with the most common of them being differential scanning calorimetry (DSC). In the following paragraphs, we discuss the use of DSC in testing the glass transition properties of epoxy coating liquids and powders.

Thermal analysis is conducted to establish the quality of epoxy powder or one or two component liquid epoxy coating materials. This test can also be carried out on the applied production coating. To conduct this test, the laboratory must use a DSC with cooling accessories, a balance that can measure to 0.1 mg accuracy, a supply of dry laboratory-analysis-grade nitrogen gas, a sample encapsulating press, and aluminum pans.

The test procedure differs for two-component and one-component epoxy liquids.

1. To conduct a one-component epoxy liquid test, 100 g of liquid is taken and homogenized in preparation for the test.
2. To conduct a two-component epoxy liquid test, both parts, the base and the hardener, are taken in the recommended ratio and mixed together. The mix is then completely homogenized in preparation for the test, and a 100-mg sample of this mixture is removed for further tests.

The epoxy powder is heated and cooled in following three ways.

1. The powder sample is heated from 25 to 70 °C. The heating rate is 20 °C/minute. As the temperature reaches 70 °C, the sample is cooled to between 20 and 30 °C.
2. The powder sample is heated from 25 to 70 °C. The heating rate is 20 °C/minute. The sample is then immediately cooled to between 20 and 30 °C, and it is held at the reduced temperature for 3 min.
3. The powder sample is heated from 25 °C to $T_g + 40$ °C. The heating rate is 20 °C/minute. The sample is then immediately cooled to between 20 and 30 °C.

Evaluation of the Result

The glass transition temperature T_g is the plotted on a graph where the temperature is plotted on the X-axis and the differential heat flow, expressed in watts per square meter, is plotted on the Y-axis. The point of inflection on the curve is the glass transition temperature T_g. In the above described process, the T_g of an *uncured* powder epoxy is obtained through the second step and referred to as T_{g1}. The T_g for both *cured* powder epoxy and liquid epoxy is obtained in the third step and is referred to as T_{g2}.

SAMPLE PREPARATION FOR COATING GLASS TRANSITION TEMPERATURE MEASUREMENT

Epoxy Powder

The applied coating is extracted from the pipe substrate. The sample should be dried before any measurements are taken. Ten milligrams of coating is removed and weighed to an accuracy of ±3 mg on a scale that can measure to an accuracy of 0.1 mg. The sealed pan bearing the sample is reweighed, and the final weight is recorded. This reference sample is then placed in the DSC cell, and air is purged with dry nitrogen gas.

One-Component Epoxy Liquid

The liquid coating sample is mixed and homogenized before being applied to an aluminum panel in a 500-μm-thick film. The aluminum panel is a 1 mm thick and cleaned before the application of coating. The coating film is allowed to cure on the aluminum for a minimum of 2 h at ambient temperature. Once the film is cured, the panel is heated to 170 °C in a ventilated oven for about 15 min and then cooled to ambient temperature. About 2 h after curing, the aluminum panel is bent, and a coating scale is scraped off to obtain sufficient mass for further testing. The same type of coating removal is done if the product is applied to a steel substrate (pipe). The removed coating is placed in the test capsule.

Two-Component Epoxy Liquid

The coating's constituent materials are mixed in the specified ratio for about 5 min before being applied to a pipe substrate or, in a laboratory setting, to an aluminum panel. The aluminum panel is 1 mm thick and cleaned prior to the application of the coating in a film with a thickness of 500 μm. The coating film is allowed to cure for a minimum of 2 h at ambient temperature. Once the film is cured, the panel is heated to 170 °C in a ventilated oven for about 15 min and then cooled to ambient temperature. About 2 h after curing, the aluminum panel is bent, and a coating scale is scraped off to obtain sufficient mass for further testing. The same type of coating removal is done if the coating is applied to a steel substrate (pipe). The removed coating is placed in the test capsule.

Measuring

After the sampling is complete, as described above, each of the three types of epoxy coatings systems can be measured. During the measurement process, a set cycle of heating and cooling is conducted for the three types of coating systems.

1. The powder sample is heated from 25 to110 °C (± 5 °C). The heating rate is 20 °C/minute. The temperature is held for 90 s, and then the sample is cooled to between 20 and 30 °C. This step is applicable to powder epoxy only. The liquid epoxy testing process starts with step 2.

2. The powder sample is heated from 25 to 275 °C (± 5 °C). The heating rate is 20 °C/minute. The sample is then immediately cooled to between 20 and 30 °C and held at this temperature for 3 min. This step provides T_{g3} for the plot.

3. The powder sample is heated from 25 °C to $T_g + 40$ °C. The heating rate is 20 °C/minute. The sample is then immediately cooled to between 20 and 30 °C. This step provides T_{g4} for the plot.

EVALUATION OF GLASS TRANSITION TEMPERATURE RESULTS

The procedure for applied coating gall temperature evaluation is similar to the procedure for determining the epoxy material's glass transition temperature. As described previously, a plot is made, and T_g is determined, with $\Delta T_g = T_{g4} - T_{g3}$ measured in degree Celsius. The exothermic heat of the reaction (ΔH_1) is determined by integrating the peak of the DSC scan, as drawn on the plot for T_{g3}. For a fully cured coating film, no residual heat reaction should be observed. The degree of conversion (C) is expressed as a percentage, and it can be determined using the following relationship.

$$C = (\Delta H - \Delta H_1) \times 100/\Delta_1$$

where ΔH is the exothermic heat of reaction of the powder, as discussed in the epoxy material evaluation above; and ΔH_1 is the exothermic heat of reaction of the powder, as discussed in the applied epoxy coating evaluation above.

TESTING FOR THE GEL TIME OF EPOXY POWDER

Gel time indicates the relative reactivity of a powder coating formulation at a specific temperature, usually expressed as the number of seconds to gelation. Using a hot plate, the gel point is indicated when stroking with a stick no longer produces a polymer thread. Accordingly, this test is commonly referred to as stroke cure gel time. The reactivity of a protective powder coating is usually measured at 205 °C and can be adjusted by the level of accelerator in the formulation for each specific application. Exterior protective powder coatings are usually separated into two categories.

1. Small diameter pipe coatings
2. Large diameter pipe coatings

These two classifications often overlap and depend on application factors such as pipe temperature, line speed, powder spray rate, desired film thickness, time before quench, and the location of pipe rollers. Generally, at

205 ± 3 °C, small diameter systems have gel times of about 5–12 s, with complete curing in 30–90 s. Larger diameter formulations have gel times of about 16–45 s, with complete curing in 60–120 s.

The equipment used for testing is:
- A hotplate that is capable of temperature control within 3 °C
- A stopwatch with a measuring accuracy of 0.1 s
- A draw tool that has the draw–end notch of about 25 mm recessed to about 1 mm deep
- A metal plate placed on the hotplate

THE TEST PROCEDURE

A set of three tests are carried out. The reported result is averaged. As a result, the gel time of epoxy powder used in the three-layer coating system is determined, according to the following procedure. The metal plate placed on the hotplate is heated to the required 205 ± 3 °C temperature. On the bottom end of the draw tool, the notch section is covered with epoxy powder, and the powder is then drawn at an approximate angle of 45° to the metal plate surface, pulled across, and deposited on the metal plate using a smooth action. Simultaneously, the timing device (stop watch) is clicked on. This action should create a draw of 25-mm-wide coating on the metal plate, with the thickness of the applied coating between 300 and 400 μm. The draw edge is repeatedly drawn through the molten epoxy until the tool starts to ride over the now gelled epoxy. At this very moment, the stopwatch is stopped, and the time recorded. The recorded time is the gel time of the tested powder epoxy.

TOTAL VOLATILE/MOISTURE CONTENT OF THE EPOXY POWDER

The powder epoxy coating is applied in the shops in a very controlled environment. However, there is still a downside to these coatings, namely contamination of air and related health risks. This risk is heightened by the lowering of air exchange rates, given that several of the chemical contaminants are trapped indoors where people spend most of their lives. The harm to clean air and thus to health is due to the volatile organic compounds (VOC) in the dust that is blown in the coating environment. The World Health Organization (WHO) defines VOCs as organic compounds with boiling points between 50 and 260 °C. The VOC measurements were conducted according to ASTM D5116-97, which was used as a guide for Small-Scale Environmental Chamber Determinations of Organic Emissions from Indoor Materials/Products. This standard was withdrawn in 2006, however, and no replacement is known to be issued by ASTM, with the withdrawn guide still in use. The rational for its continued use can be understood from the rationale for withdrawal given by the ASTM committee.

The use of small environmental test chambers to characterize the organic emissions of indoor materials and products is still evolving. Modifications and variations in equipment, testing procedures, and data analysis are made as the work in the area progresses. Until the interested parties agree upon standard testing protocols, differences in approach will occur. This guide will provide assistance by describing equipment and techniques suitable for determining organic emissions from indoor materials.

In the ASTM guide, options are described, but specific courses of action are not recommended. The purpose of this guide is to increase the user's awareness of available techniques for evaluating organic emissions from indoor materials/products via small chamber testing, and to provide information from which subsequent evaluation and standardization can be derived.

The VOC test is aimed at determining the loss of volatiles from the epoxy powder. The procedure requires following equipment:

- An oven that is capable of controlling the temperature to within 3 °C
- A balance that is capable of measuring to an accuracy of 0.001 g
- Desiccators
- A sample container

The following list describes the step-by-step procedure used to determine the VOC content of the epoxy powder.

1. Weigh the container.
2. Take a small amount of the epoxy powder sample and weigh it at 10 g with an accuracy of 0.001 g.
3. Place the container with the sample in the oven and heat it to 105 ± 3 °C for no more than 2 h.
4. Remove the container and place it on the desiccators to cool.
5. When the sample temperature reaches 23 ± 3 °C, weigh the container. This process is repeated at an interval of about 1 h until two consecutive weight readings differ by no more than 0.001 g.

The percentage moisture content is the ratio of the mass change to the initial mass of the sample. This ratio is expressed in the equation:

$$Wm = (M_I - M_F) * 100/(M_I - M_C)$$

where M_I = the initial mass of the sample container and the epoxy powder, M_F = the final mass of the sample container and the epoxy powder, and M_C = the mass of the container without the epoxy.

ULTRAVIOLET AND THERMAL AGING TEST

Many coatings, especially those containing polypropylene and polyethylene materials, are damaged by exposure to ultraviolet (UV) rays. Coatings are often placed in environments with exposure to UV rays. The test is conducted on a sample drawn from the batch and exposed to continuous irradiation from a xenon lamp in a specific environment where temperature and humidity is controlled to simulate typical conditions. After the described UV exposure, the sample is evaluated by assessing the change in its melt flow rate. For conducting this test, the following equipment is required.

- An irradiation chamber equipped with a xenon lamp. The test is regulated under ISO specification ISO 4892-2, and the details of the required irradiation chamber are described in the specification.
- A melt flow tester, as described in ASTM D1238/ISO 1133. This ASTM/ISO specification details the procedure measuring the rate of thermoplastic extrusion through an orifice at a prescribed temperature and load. The specification also provides a means of measuring the flow of a melted material, which can be used to differentiate grades, as with polyethylene, or to determine the extent of the degradation of a plastic as a result of UV-exposure molding. Degraded materials generally flow more readily as a result of reduced molecular weight, and as a result, they could exhibit reduced physical properties. Typically, a quantitative measure of flow rates for the material is determined by calculating the percentage difference from its nondegraded values to its degraded values.

TEST PROCEDURE

A set of three samples are collected for UV-degradation testing. Each sample is about 7 g, and it is loaded into the barrel of the melt flow apparatus, which has been heated to a temperature specified for the material. A weight specified for the material is applied to a plunger, and the molten material is forced through the die. At least 14 g of material is sampled through a time (t). Extrudate is collected and weighed, and the flow rate is calculated. Melt flow rate $= (600/t \times$ weight of extrudate), and melt flow rate values are reported in g/10 min. An average of the three samples is determined for the UV degradation procedure.

DENSITY OF EPOXY POWDER

The density of the epoxy powder is determined by mixing a fixed amount of epoxy powder, usually 20 g to an accuracy of 0.01 g. This sample is then mixed with mineral spirit in a flask in order to wet the epoxy powder. The sealed flask is agitated for several minutes to ensure the removal of any lumps or air pockets from the mixed powder. The flask is then filled with the mineral spirit to the 100-ml level and weighed to a 0.01-g accuracy. The weight is recorded, and the flask is emptied, cleaned, and dried. The dried flask is filled with 100 ml of mineral spirit and weighed to the same accuracy of 0.01 g.

Calculation of densities:

The density of mineral spirit ρ_s is measured in grams per liter and calculated using the following formula, where the mass of the flask with mineral spirit is expressed as M_{fs} and the mass of the flask alone is expressed as M_f.

$$\rho_s = (M_{fs} - M_f)/0.1$$

The density of the epoxy powder is similarly calculated using the following formula.

$$\rho_p = (M_{fp} - M_f)/0.1 \left\{ (M_{fps} - M_{fp})/\rho_{ps} \right\}$$

where ρ_{ps} is the density of the mineral spirit in g/l, M_{fp} is the mass of the flask with epoxy powder in grams, M_f is the mass of the flask alone in grams, and M_{fps} is the mass of the flask with the mixture of epoxy powder and mineral spirit in grams.

PRODUCTION TESTS

After the coating is applied to the substrate, several inspection and test methods can be used to verify the quality of the applied coating. These tests are most often verify the physical properties of the applied coating, including coating flexibility, and resistance to impact-related damage. The testing of the applied coating determines the thickness and bonding prorates of the coating, and testing at this stage also validates the properties of the coating material and the procedure of application.

Inspection of Thickness

Measuring coating thickness validates the coating application procedure and the coating material.

The measurement is carried out using a calibrated thickness-measuring gauge that is accurate to within 0.1 of the measuring scale. The gauge is calibrated at the start of the day, and its accuracy is verified by measuring a thickness that is within 20% of the thickness to be measured. During the production application of the coating, the pipe is checked at 12 locations distributed along the length and circumference of the pipe. All readings should show the specified thickness of the coating. Additional measurements of thickness on the weld seams are often specified in the requirements. All measurements taken are reported in the final inspection reports.

HOLIDAY TEST

A Holiday test is carried out to determine the presence and location of any electrical flow in the applied coating. The discontinuities, such as porosity, that are open to the steel substrate are detected using an electrode charged with high arc-voltage. The electrode is swept over the surface, and any leakage to the substrate breaks the flow of the current reaching the steel substrate, producing an arc spark. This spark is often supported by a high-pitch alarm indicating the location and size of the defect. Voltage on the equipment is set to 10 kN/mm, based on the minimum total coating thickness. The total voltage does not exceed 25 kV. The scanning electrode is passed over the surface of the coated structure or pipe so that it covers the full surface. The speed of scanning movement is slow enough that a 1-mm diameter defect can be effectively picked up by the scan.

PEEL STRENGTH TEST

This test is conducted to ensure the adhesion of the coating to the substrate. Accordingly, the test measures the strength required to peel the applied coating off of the substrate. To conduct the procedure, a tensile testing machine is used. This machine should be capable of recording the peel force with an accuracy of 5%, and the rate of pull is set to 10 mm/minute. The length of peel and the temperature at which the peel test is performed is often specified in technical specifications. On pipes, the test is carried out on a ring of coating strip cut on the pipe coating in the radial direction. This strip is gripped in the jaws of the peeler, while the strength required peel the coating is recorded.

IMPACT TEST

The impact test determines the toughness of the coating, thus measuring the coating's ability to resist puncture or damage from the impact of a citrine load. The test is carried out at a given temperature, and the impact load is dropped from a specific distance to achieve the desired velocity and associated impact energy. To conduct this test, a drop weight testing (DWT) machine is used with the test sample. Standard DWT machines are equipped with support and leveling devices, and height-measuring tools are often attached to the main frame of the machine, or a graduated ruler is used. The machine is loaded with a 25-mm diameter steel punch that has hemispherical head. This weighted punch is dropped from a height onto the coated pipe specimen that is often maintained at a temperature of 23 ± 3 °C. After the impact, a holiday test is carried out to detect any opening in the coating. For accurate test results, the pipe is always positioned perpendicular to the center line of the drop punch.

HOT WATER IMMERSION TEST

This test is carried out on factory-applied three-layer polyolefin coatings to determine the coatings' ability to resist the loss of adhesion to the steel substrate. This test determines the ability of the applied coating to withstand the wet environment. The hot water immersion test is often part of the coating procedure qualification tests, and it is carried out on a sample cut from a pipe that is coated according to the specified procedure. The sample is prepared for immersion in a hot water bath maintained at about 80 ± 3 °C. For the test, a 150-mm ring is cut from small bore pipes or a 150×100 mm sample is removed from large bore pipes. A set of three specimens are then prepared by grinding the exposed surfaces with 120-grit sand paper.

In a glass beaker of suitable size, distilled or deionized water is heated to 80 ± 3 °C, and the prepared specimen is immersed in the beaker such that about 50 mm of water is above the immersed specimen level. The specimen is kept under these conditions for 48 h. After 48 h, the specimen is removed and dried with towels, and the sample is examined at room temperature. Visual examination is conducted to determine if there is any adhesion loss at the coating-substrate interface. Ignoring the loss appearing within 5 mm of the edges, if a loss is noted, it is further investigated by lifting the opening along the longitudinal axis of the pipe with a sharp knife tip. The objective is to assess the length of the opening. The maximum depth of the adhesion loss shall be recorded, and if more than one such location is recorded in the set of three specimens, an average depth is calculated and recorded.

INDENTATION TEST

This test determines the ability of the coating to resist damage caused by a sharp object impinging on the coating for some time. A temperature is also set as factor of the testing. The penetrometer used in the test is equipped with a sharp object 1.8 mm in diameter, called an indenter. The indenter is loaded with a 2.5-kg weight to produce a force of 25 Newton. A dial gauge is used to measure the depth of the indent to 0.01 mm accuracy. The pin is pressed to the specimen in a ventilated chamber where the temperature is maintained within 2 °C of the specified test temperature. The specimens are held at the temperature for an hour, and the reading on the dial gauge is recorded. At this point the specimens are loaded with the2.5 kg weight and kept in position for 24 h. After the test time, the dial reading is recorded again. The average of the final readings for the three specimens is calculated and compared with the initial gauge reading. The change in the two readings is the indentation value of the coating.

CATHODIC DISBONDMENT TEST

The coating-applied structure is often subject cathodic protection that has a specified polarization current applied to it, or, if the structure itself is not protected by a CP system, it may share an electrolyte with a structure receiving protective polarization current. If this polarization current is high and thus capable of producing a significant amount of hydrogen ions over the surface the cathode, then the pressure of the produced hydrogen can disbond the coating from the steel substrate. This disbonding can increase if alkaline ions are increased during the cathodic polarization process. The cathodic disbondment test is carried out at various polarization current levels and temperatures that would give the following polarization potentials.

a. −1.5 V, 20 ± 3 °C for 28 days

b. −3.5 V, 65 ± 3 °C for 24 h

c. −1.5 V, maximum operation temperature limited by the coating system's temperature limits (often limited to 90 °C), for 28 days

The test for cathodic disbondment is one of the key methods for determining the efficiency of a good coating. The specimen is cut from a holiday-tested section of the pipe, and a defect of a specific size, often 6 mm in diameter, is drilled in to the specimen. The drilled defect penetrates through the coating layers and reaches the steel substrate. The test requires following equipment:

1. A 100×100 mm specimen from the coated pipe
2. A rectified DC power source with voltage output control
3. A calomel reference electrode
4. A hotplate with temperature control to the accuracy of ±3 °C
5. A steel tray filled with sand at the required temperature
6. A platinum wire electrode (0.8–1.00 mm)
7. A 75-mm diameter plastic cylinder
8. 3% Sodium chloride (NaCl) solution in distilled water
9. A utility knife

The test preparation includes attaching the plastic cylinder over the artificial defect created by drilling the test specimen. The cylinder is centered over the defect and then attached to it via a waterproof sealant. The cylinder is then filled to about 70 mm with the NaCl solution, which is pre heated to the test temperature, and the level of the solution is marked on the cylinder for future reference. An electrode is inserted through the top cover of the cylinder and connected to the positive terminal of the DC power source. The negative terminal is attached to the set voltage for test, which is negative to the Calomel reference electrode. During the test period, the level of NaCl solution is maintained by adding fresh NaCl solution to the cylinder.

After the test time is reached, the test apparatus is dismantled, and the specimen is removed and cooled in air. Cathodic disbandment specimen is evaluated within an hour of removal from the heat. The evaluation involves making 6 to 12 radial cuts with the knife, starting from center of the drilled defect to about 20 mm. The cut should be deep enough to cut the coating down to the steel substrate. Once the cuts are made, the coating is chipped off with the tip of the knife, using a lever action, until there is resistance to the chipping. The disbondment distance in each direction is measured, and an average is calculated. The average disbondment in mm is the coating's cathodic disbondment.

FLEXIBILITY TEST

As the name suggests, this test establishes the flexibility of a three-layer application of polyolefin coating. A specimen of specific size, generally about 200 mm long and about 25 mm wide, is cut from the coated pipe. For this test, apart from the thickness of the steel substrate and coating, a new term, effective thickness (d), is used. The effective thickness is the two above-described thicknesses added to the measure of the curvature that arises from specimen being cut from the pipe section. This total thickness is called the effective thickness for the test. This effective thickness (d) is used to determine the maximum radius of the mandrel on which the specimen will be bent. The radius (r) corresponds with the 2° per pipe diameter length. The value of (r) is determined as follows.

$$r = 28.15 \times d$$

The specimens are soaked for 1 h at −2 and 0 °C, generally in a freezer. Within 30 s after the specimen's removal from the soak, the specimen is bent over the mandrel. The bending operation should be completed within 10 s. After the bend, the sample is allowed to warm up to an ambient temperature of 23 °C and to soak in a fluid at that temperature for about 2 h. Once the sample has attained the ambient temperature, it is visually examined for the presence of any cracks. Any appearance of a crack will constitute failure of the specimen and the coating.

SECTION 5

Maintenance of CP System and Retrofitting to Continue Cathodic Protection

CHAPTER ELEVEN

Introduction to Retrofitting

The previous chapters extensively discuss the corrosion protection of off-shore structures. That discussion primarily focuses on the design of new structures and the efforts required to protect those new structures from corrosion damages and failures. This discussion of corrosion protection includes a consideration of system designs that use the corrosion protection properties of coatings in combination with cathodic protection (CP) methods. In earlier chapters, we also point out that, in general, both coating and CP systems are used together so that each method complements the exclusive properties of the other.

In this chapter and subsequent discussions, we focus on a new trend in corrosion protection that represents a very logical next step in safeguarding offshore structures from electrochemical deterioration. That trend is the provision of additional CP protection or the extension of existing protection for structures that have exceeded their initial CP design life. Many of these structures, whether platforms or pipelines, would require a retrofit of the CP system during their lifetimes. The retrofit would most likely occur at the end of the designed life time, but such is not necessarily the case. In most cases, a retrofit is called for in two primary situations.

1. The initial CP system is unable to provide sufficient current output to protect the structure.
2. The operator has decided to extend the use of the structure beyond the initial design life.

The second point is clear in the sense that operators are expecting more output from their existing structures, as they want to stretch the value of their investment. However, the first point is more the cause of the retrofit, in that the evaluation of currently installed conditions of anodes, their depletion rates and thus the current output are on one side and the provision of new set of anodes calculations and distribution of current become the essentials of new design activity.

Often a combination of both conditions is the reason for considering a retrofit of the CP system. So, when a retrofit is needed, all the discussed principles of CP design and related calculations come into play. However,

Corrosion Control for Offshore Structures
165

several design constraints are specific to a new CP design, such as the use of a bracelet anode for a pipeline being installed or the possibility of anode loss due to the force of installation or platform piling.

Structures require retrofits more frequently than normally thought, because of the fact that production and transportation systems are getting older worldwide. Many offshore pipelines transporting oil and gas have been in service for more than 50 years now, and in some cases, even longer. Most of these pipelines are protected from external corrosion by a combination of coatings and CP, as we discuss in pervious chapters. A majority of the older pipelines and structures are operating past their initial design lives, however, and hence, operators must evaluate the current protection provided to these structures and pipelines, while considering a possible retrofit of the CP system.

For example, the U.S. Gulf of Mexico now contains ∼30,000 miles of crude oil and gas pipelines, and as with similar structures around the world, the ages of these pipelines vary, leading to different retrofit requirements. Occasional accidents and failures do occur, but the pipelines in the Gulf of Mexico are generally considered safe, efficient, economically effective transportation methods. As a result, industry relies on these lines to move offshore oil and gas from fixed production facilities. As safe as they are, these pipelines occasionally fail, however, causing environmental damages that sometimes involve the loss of human lives. Most of these failures can be grouped as failures of:

- Material and equipment
- Operational errors
- Corrosion
- Storm damages
- Third party incidents (mechanical damage)

As stated above, these failures can be responsible for loss of life, environmental pollution, loss of product availability, repair expenses, business interruption, litigation, damage, and loss of reputation. Several current publications list these failures worldwide, and some of the publications have evaluated these occurrences and analyzed the causes of the involved offshore pipeline failures.

These failures can involve both pipelines and risers, but a significant percentage of these failures have occurred on pipelines as opposed to risers. Most of the failures are due to external corrosion, and very few are associated with internal corrosion. This difference in the distribution of corrosion indicates the pipe material's susceptibility to corrosion when it is in contact with the water's surface where the corrosion rate is generally the greatest. In fact, a significant number of these failures are detected prior to substantial product discharge, and at that point, it is relatively inexpensive to repair them. Even when identified early, corrosive damage is not to be taken lightly, however, as the easily seen damage probably underrepresents the effect of corrosion on the pipeline. In many instances, a pipeline that was initially weakened by corrosion later failed due to an alternative cause such as a storm or third party damage. In such cases, pipe failure is often attributed to the alternative cause and not to the corrosion. Other concerns with regard to pipeline corrosion failures can be as described in the following data from the Gulf of Mexico. Similar data may be available or at least can be collected for other parts of the world where such failures have occurred. This collected data can then be analyzed.

- The average failure rate during the 1990s was more than double the failure rate in the 1980s,
- The increased focus upon deep water production in the Gulf of Mexico indicates that failures, where they occur, will be more difficult and expensive to address. Worldwide, other fields have faced similar challenges, given the move toward deeper drilling and production.
- For many older pipelines, the CP system's design life has now been exceeded, such that external corrosion may be ongoing and CP retrofitting may be required.

Given the absence of robust design procedures for retrofitting the CP systems on existing offshore pipelines, the process of CP retrofitting has remained unstandardized. In recent years, increased attention has been directed toward the external corrosion of offshore pipelines, however. The corrosion data and analysis discussed above may include the following steps for identifying and addressing CP retrofit design challenges.

- Verify the proposed CP design method and attenuation model.
- Study and analyze the retrofit design process for offshore pipelines.
- Adopt an inclusive first-principles-based attenuation model for offshore pipeline CP.
- Define and examine critical issues related to pipeline CP retrofits.
- Review and revise the CP design process for new pipelines.

A critical review of the existing CP design process for offshore structures should be conducted. This review should concentrate on the protection of various structures exposed to seawaters, including the CP of platforms, pipelines, and other structures. The current design slope and drain calculations should also be performed. In the following paragraphs, we briefly discuss corrosion and its impacts. This discussion somewhat repeats our previous discussion, but it is necessary to keep the subject of retrofitting anchored to the reasons for considering retrofitting. In considering the need to prevent the negative impacts of corrosion, one should remember that pipeline corrosion inspections are often neither sufficiently sensitive nor sufficiently comprehensive to necessarily disclose problems, and pipelines may experience modified service conditions. This requires that the retrofitting criteria are defined, and an acceptable industry standard is established. The following issues relating to offshore pipelines must also be investigated and addressed.

• Consider a new approach to CP design of offshore pipelines is considered.
• Develop an inclusive, first-principles-based attenuation model for offshore pipeline CP.
• Develop a CP design method and attenuation model
• Identify and define critical issues related to offshore pipeline CP retrofits.

The retrofit CP design process involves the steps listed below, and these steps are based on the general principles of CP discussed in the previous chapters.

Step 1: Assess the conditions to determine if a retrofit is actually required.
Step 2: Calculate the required current density. This will help establish the required anode output.
Step 3: Establish the determining factor for anode array design. Is it based on its proximity to a structure or it is based on minimization of the anode battery?
Step 4: Design an anode battery.

ECONOMICS OF RETROFIT

The cost of retrofitting pipelines and platforms is often considered during the design process. Industry has analyzed this cost, and there are several published reports that can serve as references for budgeting and estimation. The objective of developing cost projections for the retrofit should be to balance the installation cost with the integrity of the installed system. It may be noted that the installation cost is the most expensive part of the retrofitting. There are many ways the retrofit cost can be trimmed to meet the specific conditions of installation, and some of these cost-cutting measures are listed below.

1. Reduce the number of anode installation locations.
2. Reduce the bottom-time required for each location.
3. Compare the costs of moored systems and dynamically positioned system.
4. Consider the cost of using a ROV over the cost of using divers.
5. Select a pipeline location that is exposed and not buried, because more current is required for a buried pipeline section.
6. Where possible consider a mix of ICCP and GACP systems.
7. Conduct ongoing periodic surveys to tailor the current demands.
8. Reduce marine growth removal during installation on structures.

The periodic survey mentioned above can make a very significant impact on the design and cost of the system being installed at different depths and locations (see Appendix table).

CHAPTER TWELVE

Corrosion Control for Existing Offshore Pipelines

High-strength steels are commonly used in the construction of offshore oil and gas transport pipelines, platforms, and processing facilities. As a result, such structures face an associated risk of corrosion due to the lack of corrosion resistance in this type of steel, especially in seawater. The low corrosion resistance of high-strength steel then results in a potential for corrosion failure. Given the risks, the need for corrosion control cannot be overemphasized, and it demands a robust corrosion control system designed, installed, and maintained such that a high degree of reliability is achieved.

Historically, cathodic protection has been employed as the sole control methodology for the submerged portions of petroleum production platforms, both mobile and fixed, but, given the one-dimensional nature of pipelines, these structures tend to require the combined use of coatings and CP. As we have discussed for structures such as platforms, anode resistance, structure, and current density demand are the fundamental parameters determining CP design. For pipelines, however, coating quality and metallic path (pipeline) resistance are also taken into consideration.

CP systems can be designed using either the impressed current (IC) or galvanic sacrificial anode current (GA). This is true for both pipelines and other structures. For offshore pipelines, the application of IC is limited by the proximity of the pipeline to the shore. This limitation is due to the fact that a shore-based rectifier and anode array can only provide seaward protection up to a distance defined by the "throwing power" of the system. Impressed current cathodic protection (ICCP) could also be used to protect a pipeline when that pipeline is short and runs between two platforms. The choice of IC depends on whether the length of the pipeline can be protected by using a single rectifier and anode, or an anode template at one or both ends. In either of the two cases, the range of corrosion protection is normally limited by the voltage drop along the metallic pipeline that arises in conjunction with the current return to ground. The quality of the protective coating also affects the distance to which corrosion protection is afforded. Thus, the higher the coating quality, the less the pipe current demand, and, as a consequence, the less the voltage drop for a pipeline of a given length. The coating quality of offshore pipelines is normally lower than the coating quality of onshore pipelines buried under ground, however. As a result, coating breakdown factors affect the efficiency of a coating as it deteriorates over time, and the range of corrosion protection is, in turn, reduced.

In order to maximize the distance to which protection is achieved, operators tend to raise the potential within the system. This increased potential overprotects the region of the pipeline near the rectifier and anode array. Such overprotection can cause coating damage in the form of blistering and disbondment, in which case the pipe current demand increases. Because of these factors, corrosion control for the great majority of marine pipelines is provided by sacrificial galvanic anodes (GACP), and due to structural, economic, and installation considerations, these GACP anodes are invariably of the bracelet type, as shown below in Figure 12.1. These structural and installation considerations limit the size and weight of bracelet anodes and thus the distribution of anodes on the structure. In most cases anode spacing does not exceed 750 ft. (250 m). This spacing leads to the factor of voltage drop that occurs in the pipeline, and the CP system's life is governed by anode mass considerations alone.

AGING OFFSHORE PIPELINES

CP systems for offshore pipelines are typically designed for 20–25 years design life. The CP design life for many of the early pipelines has already been exceeded, but the older pipes are still in service.

Cathodic protection designs were typically less conservative in the past, as compared to current practice, and this difference in CP systems is further

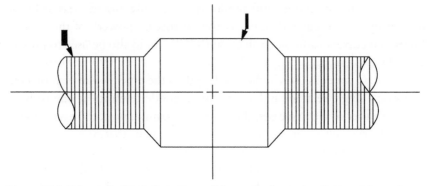

Figure 12.1 Schematic illustration of an offshore pipeline galvanic bracelet anode.

affected by improvements in anode quality over time. However, regardless of the system's date of manufacture, it is not uncommon for the electrical connection between an anode and a pipeline to become damaged or disengaged, or for an anode to become otherwise compromised during pipeline installation, and any of these occurrences could effectively double the anode spacing. Poor anode quality implies that the anode output is not what it is supposed to be. This could be due to faulty chemistry or internal inclusions that lead to either or both the following.

1. The pipeline would fail to polarize.
2. The anode is consumed much earlier than its designed life.

As a result of these two conditions associated with poor anode quality, the external corrosion protection systems on older pipelines fail to perform as designed. These failures lead to pipelines' cathodic protection systems requiring upgrading, and the pipelines' CP systems must be retrofitted. Thus, retrofitting is increasingly common and more frequently applied.

CATHODIC PROTECTION DESIGN FOR OFFSHORE STRUCTURES

In previous chapters, we provide detailed explanations of the design requirements and criteria for cathodic corrosion protection in offshore structures. We know that the corrosion of steel in seawater is arrested by polarization to a potential of $-0.80 \, V_{Ag/AgCl}$. So, the main goal of cathodic protection is to achieve and maintain a minimum polarization based upon this potential, irrespective of the type of structure being protected. The pipeline is most polarized in the regions directly surrounding the anodes, and potential attenuates as the distance from the anodes increases. Four factors determine the magnitude of this potential attenuation, as listed below.

Anode Resistance

Resistance is encountered as current leaves the anode and enters the electrolyte (seawater). This resistance is a consequence of geometrical confinement in the vicinity of the anode. The attenuation caused is greatest in magnitude immediately adjacent to the anode and decreases with increasing distance.

Coating Resistance

The objective of coating is to reduce the exposed surface area of the pipeline, thereby enhancing the effectiveness, efficiency, and distance to which corrosion protection is achieved. The intrinsic resistivity of marine pipeline coatings is relatively high, but nearly all coatings are affected by defects and bare areas resulting from handling, transportation, and installation. Consequently, the CP current enters the pipe at these locations where steel is directly exposed.

Polarization Resistance

The term polar resistance reflects an inherent resistance associated with the cathodic electrochemical reaction, whereby ionic current in the electrolyte is translated through electronic conduction in the pipeline.

Metallic Resistance

Although the resistivity of steel is orders of magnitude smaller than the resistivity of seawater, the confined pipeline cross-sectional area, combined with the relatively long distance (length of pipeline) that the current may have to travel in returning to electrical ground, creates resistance in the metal. This metallic resistance may not be the controlling factor, but it invariably influences certain conditions and must be considered.

Portions of a pipeline for which the potential is -0.80 $V_{Ag/AgCl}$ or more negative are protected, whereas locations where the potential is more positive are unprotected. The cathodic protection design protocols specific to the platforms and pipelines are discussed below. We have noted that the cathodic protection design procedures for platforms have evolved historically according to the following progression:

- Trial and error
- Ohm's law employing a single, long-term current density
- Ohm's law and rapid polarization employing three design current densities, an initial (i_o), mean, (i_m), and final (i_f)
- The slope parameter method

The application of long-term current density and rapid polarization current densities is derived from the following derivation of Ohm's law

$$I_a = (\emptyset_c - \emptyset_a)/R_a$$

where I_a is the individual anode current output, $\emptyset_c$ is the closed circuit cathode potential, $\emptyset_a$ is the closed circuit anode potential, R_a is the resistance of an individual anode.

Anode resistance is normally the dominant component of the total circuit resistance for space-frame structures, such as platforms, hence, anode resistance receives more consideration in the subsequent discussion. In most cases, this parameter is calculated from standard, numerical relationships that are available in the published literatures and are based on anode dimensions and electrolyte resistivity.

Considering that the net current for protection (I_c) is the product of the structure current density demand (i_c) and surface area (A_c), the number of anodes required for protection (N) is determined from the relationship

$$N = (I_c \times A_c)/I_a$$

In the past CP designs were primarily based upon a single, time-averaged or mean current density required to polarize the structure to the required protection potential of $-0.80\, V_{Ag/AgCl}$ within perhaps several months. Experience taught designers that initial high current density would provide quicker polarization, and mean current could be lowered for the maintenance of the cathodic protection system. As discussed in the earlier sections, current CP design practices are based upon three current densities, initial (i_o), mean (i_m), and final (i_f), where the first is relatively high to attain rapid polarization, the second is the time-averaged value, and the third reflects what is required near the end of the designed life of the structure, when the anodes have depleted, and the current output of the individual anode is significantly reduced. The required number of anodes corresponds to the respective values for initial current density (i_o) and final current density (i_f), and this number is determined by substituting each of the parameters for structure current density (i_c) in the equation below. The mean current density (i_m) is then calculated from the mass balance relationship,

$$N = (8760 \times i_m A_c T)/uCw$$

In the above relationship:
T is the design life of the CP system (years).
u is the utilization factor.
C is the anode current capacity (A h/kg).
w is the weight of an individual anode (kg).
The number of anodes determined according to each of the three calculations is different. So, choosing the correct number requires some optimization calculations, and in most cases, the largest value is often the specified number of nodes. For uncoated structures, this is usually i_o, implying that the CP system is normally overdesigned in terms of the other two current densities. This overdesign results from the number-determining procedure being an algorithm rather than being first-principles-based.

Often the slope parameter approach to galvanic cathodic protection design is based on the following relationship between the total circuit resistance. The equation identifies the linear interdependence between the close circuit potential and current demand, on the condition that the total resistance, surface area, and close circuit anode potential are held constant.

$$\emptyset_c = (R_t A_c) \times (i_c + \emptyset_a)$$

• For the space-frame type of structures with multiple galvanic anodes,

$$R_t = R_a/N$$

• Where the product of R_t and A_c in the above equation is defined as the slope parameter (S), such that,

$$S = (R_a A_c)/N$$

• The substitution of the latter expression into the above equation gives the following,

$$R_a w = (i_m TS)/C$$

In this unified-design equation approach, all terms on the right side of the equation are determined by the design parameters, once the appropriate value for S is known. An anode type is then either selected or designed based upon the optimum combination of R_a and w. Type selection involves choosing from among an array of commercially available anodes and from among the available standard dimensions, with the possibility of special orders. The second option is limited by order quantities, but for a major project, this option is not only possible, it is also very much desired to optimize the size and cost impact on the project. As described in earlier chapters, application of Dwight's equation also describes the above relationship for anode resistance. This can be expressed as given below.

$$R_a = (\rho_e/2\pi l)[\ln(4l/r) - 1]$$

In the above relationship the ρ_e is the resistivity of the seawater (the electrolyte), the l is the anode length, and r represents the equivalent anode radius. Through further derivation, with the addition of the anode density (ρ') and the volume fraction of the anode metal minus the core (v), the previous equation is changed to the following,

$$R_a w = (\rho \rho' v)/2[\ln(4l/r) - 1]$$

This new equation allows for the determination of the total required number of anodes, using the following equation, as described above.

$$S = (R_a A_c)/N$$

As expressed in the paper by Hartt et al., various researchers have established that the slope-parameter-based design approach reduces the total mass required by a significant percentage, when compared to conventional design practices. This reduction in the mass requirement is due to the first-principles-based design that incorporates both i_m and i_o, as used in the slope parameters. This design approach then optimizes both parameters.

PIPELINES

As we have pointed out in earlier chapters, there are two fundamental differences between the designs of cathodic protection systems for offshore platforms and pipelines. These design differences exist despite the fact that same three design current densities, i_o, i_m, and i_f, are employed. These differences are due, in part, to the geometry of the structures, as platforms have three-dimensional structures, while pipelines have nominally one-dimensional profiles. In addition, pipelines are coated, while a significant section of the offshore platform is either not coated or is coated with less efficient coating systems.

However, the CP design of offshore pipeline considers the current demand, I_c, as

$$I_c = A_c f_c i_c$$

As we know, the f_c is the fraction of the holidays in the pipe coating that are exposed substrate areas. The design values for i_o and i_f are normally taken to be between 5.6 and -20 mA/ft^2, and the variation in the values depends on factors such as the depth of the pipe lay, temperature, whether the pipe is in seawater or mud, and the basis of current determination for the initial (I_c) or final (I_f) condition. This calculation of the design values for pipelines is contrasted with the design calculation for structures such as platforms, where the design is accomplished by substituting i_o and i_f for i_c and calculating the corresponding I_c. The net anode mass (M), on the other hand, is determined from the following modified equation.

$$M = (8760 \times I_m T)/(uC)$$

In the above equation, I_m is the mean current required to protect the pipeline during the design life of the pipeline. The required number of anodes is then determined by considering the values for I_o, I_f, and M. For the design of the pipeline's cathodic protection system, the permissible anode bracelet size is determined by structural and installation considerations. This due to the fact that bracelet anodes are relatively small, and they are spaced closely. As a result, metallic path resistance and voltage drop are negligible and thus rarely considered in the design calculations.

CHAPTER THIRTEEN

Corrosion and CP System Assessment Methods

An effective retrofit design begins with an assessment of the current CP system. Determining the corrosion state of a marine pipeline is difficult, however, and this difficulty often leads to the pipeline becoming underprotected because of inadequate or expired CP system anodes. This situation differs significantly from the situation faced by CP systems on platforms, because, in the latter case, a simple drop-cell-method potential survey can be performed annually or at other regular intervals, with the annual surveys complimented by more detailed surveys at 5-year intervals. The space-frame nature of platforms also ensures that a section of the structure receives protection from multiple anodes.

Some pipelines are protected by a single galvanic anode that is designed and positioned to safeguard a specific length of the pipeline. Facilities' cathodic protection systems also impress some of their current on the sections of pipeline in their vicinity, potentially causing false protection readings for those pipes. As a result, operators can often develop an inaccurate evaluation of a pipeline's CP protection, if readings are only collected at points of convenience, such as the platform of pump stations. The need for more accurate current assessment has led to the development of methods that could bridge this gap in the measuring system. Some of the survey methods used to assess over-the-line corrosion and adequate CP coverage are listed below.

1. Towed vehicle/trailing wire potential measurements
2. ROV-assisted remote electrode potential measurements
3. ROV-assisted/trailing wire potential measurements
4. Electric field gradient measurements

The towed vehicle and remotely operated vehicle (ROV) methods are based upon pipeline potential measurements that are made either continuously or at closely spaced intervals along the pipeline, using a reference electrode mounted on the towed vehicle or ROV. The advantage of using an ROV is that it allows for visual imaging, and the electrode can be placed close to the pipeline at unburied pipeline locations. In the case of the towed vehicle arrangement, the buried pipeline location must be assumed, and an assumption is also made about whether the reference electrode is "remote" or "semi-remote" relative to the pipeline, in which case only potential variations from long-line effects are disclosed. This arrangement restricts the assessment of localized corrosion spots. Figure 13.1 is a schematic illustration of a pipeline with a galvanic anode and the resultant potential profile.

As we note, the potential is relatively negative at the anode and positive at the coating defect(s) on the pipeline. However, the measured profile becomes relatively flat as the distance between the reference electrode and the pipeline increases. This relative flatness in the profile is the reason that the remote electrode will not identify the presence of coating defects and any associated localized underprotection.

The flattening of the profile does not occur with the electric field gradient measurement method, however. This method utilizes two or more electrodes and is based upon the potential difference between these electrodes. The recorded potential difference points to the current that either flows into the pipeline or outward from the anodes. If the electrodes used for the measurement can be positioned close to the pipeline, then this method is more sensitive than the other three methods described above. Thus, it can provide more accurate values to be used in determining the location and severity of coating defects. It can also help operators accurately determine the current output of anode.

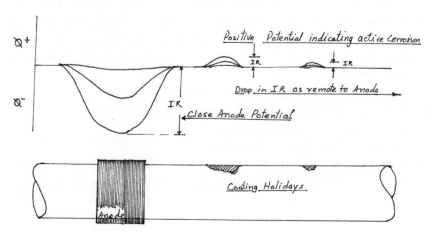

Figure 13.1 Schematic illustration of a pipeline and the potential that is recorded for an over-the-line survey.

EXISTING PIPELINE CATHODIC PROTECTION SYSTEM ANALYSIS METHODS

Previously, we have discussed the use of Morgan and Uhlig's equations to project the potential attenuation along a pipeline and also to determine the current required to achieve the indicated level of polarization in terms of coating properties, pipe properties, and pipe dimensions. Through those equations, the evaluator establishes that the current terms associated with a CP anode conform to the model of cathodically polarized pipe. The pipe polarization value is E_b, and the polarization value at specific distance z is depicted by E_z. The concept then leads to a differential equation which is resolved as,

$$E_c(z) = E_b \operatorname{Cos} h\alpha'(L-z)$$

and

$$E_a = E_b \operatorname{Cosh}(\alpha' - z)$$

In the above equation, the value of α' is the attenuation constant, and it equals the square root of the ratio of metallic pipeline to coating resistance per unit length (R and z, respectively). As shown in this equation, the CP design for an existing pipeline is able to generate adequate current output, achieving the polarization needed to protect the pipeline. This also validates the anode mass required for the protection of pipeline. The current output Ia of a single anode is a function of the pipe radius ($r\rho$), an attenuation constant based on the product of the pipe's metallic path resistance and coating resistance per unit length ($\sqrt{R_m/\zeta}$), pipe polarization value (E_b), and the relation of half-spacing of anode positioning (L), and the calculation that combines these variables and constant is based on the following equation.

$$I_a = 2[2/r\rho]\alpha' E_b \operatorname{Sinh}(\alpha' L)$$

This equation is further refined to include the polarization characteristics of bare metal exposed at the base of coating defects and the effective coating resistivity. Uhlig used the equation to calculate the anode current output; the modified equation is given below.

$$I_z = [2E_b/R_m] \times [2\pi r\rho/k\zeta]^{0.05} \operatorname{Sin} h[(z/2) \times (2\pi r\rho/k\zeta)]^{0.05}$$

In the above modified equation, the value of the pipe potential E_b at distance z is equal to L, which is the half-spacing distance of anode positioning.

The above equation acknowledges the corrosion control role of high-performance coating systems and gives it credit. This inclusion of coating effects reduces the anode mass required. So, when a good quality coating system is used, the potential attenuation (z) equals the one-half anode spacing position (L), where potential should positive enough to be minimal, in the range of about a few millivolts or less. This is possible if the anode spacing is not excessively large. But as the age of the anode increases, the current output reduces, and the consequent attenuation increases. Coating deterioration also contributes to increased attenuation. As a result of these discussed factors, alone or in combination, the current demand of the pipeline exceeds the anode output capability. The onset of such potential attenuation is a fundamental indicator that a need for CP retrofit is eminent, even though the pipeline may still be protected, with E_z values more negative than $-0.80\ V_{Ag/AgCl}$ all along the pipeline.

A key limitation of the above-discussed calculation is the fact that current output is not considered as a function of anode resistance. Thus, this approach does not address optimization of anode spacing or the evaluation of anode depletion and the onset of underprotection, except in the special case where anode resistance is negligible. For the pipeline, this transition from protection to underprotection occurs during a relatively short time frame, as compared to the transition for jackets and platforms.

Boundary-element modeling is used to conduct the analysis of potential attenuation along length of pipelines, as well as anode current output. This approach incorporates the electrolyte and coating resistance terms and uses the numerical algorithm for the solution of a Laplace-type governing equation. The equation describes the potential variation in an electrolyte. To model an electrochemical process, the Laplace equation is used in conjunction with specified boundary conditions that portray the geometry and effects of electrical sources and sinks. The limitation of the process is that it excludes the metallic pipe path component.

CHAPTER FOURTEEN

New Approach to Cathodic Protection Design for New Pipelines

INTRODUCTION

In previous chapters, we discuss the use of the slope parameter approach in designing CP systems for platforms and other offshore space-frame structures. Now, we wish to determine whether the slope parameter approach can be applied to pipelines. In this regard, application of equations for a coated and cathodically polarized pipeline requires following:

- The spacing between anodes must be small enough that the metallic path resistance is negligible.
- The pipe-to-seawater resistance is negligible.
- All current enters the pipe at holidays in the coating (bare areas).
- f_c and f_a are held constant despite changes in time and position.

Given the above-listed conditions, the total cathodic protection circuit resistance is in congruence with the individual anode resistance, a relationship expressed as $R_t \cong R_a$. The pipe surface area protected by a single anode is $A_{c(1)}$. The ratio of the total pipe surface area to the bare surface area is γ, and the 2L anode-spacing is expressed as L_{as}. The value of γ is the modified way of describing the coating breakdown factor f_c, as employed in the earlier calculations. Using the described variables, the following equation determines $A_{c(1)}$.

$$A_{c(1)} = (2\pi r \rho L_{as})/\gamma$$

If we combine the above equation with the equation in Chapter 12 about the linear interdependence relationship between $\varnothing_c$ and i_c is given by the function of total circuit resistance R_t, A_c, and constant $\varnothing_a$. The resulting equation describes the relationship in terms of L_{as}.

$$L_{as} = [(\varnothing_c - \varnothing_a)\gamma]/(2\pi r \rho R_a i_c)$$

This linear interdependence can be further expressed in terms of cathode current demand (i_c) as the ratio of the difference between the free corrosion potential and the polarization resistance. This relation is expressed using the following equation.

$$i_c = (\varnothing_{corr} - \varnothing_c)/\alpha$$

If we insert this value of i_c in the above L_{as} equation, the L_{as} equation can be rewritten as

$$L_{as} = [(\varnothing_c - \varnothing_a)/\varnothing_{corr} - \varnothing_c] \times [(\alpha\gamma)/(2\pi r\rho R_a)]$$

This relationship can be graphically expressed as shown in Figure 14.1, where current density is plotted on the X-axis and the corrosion potential is on Y-axis. The corresponding design life (T) in years is then calculated using following equation.

$$T = (wCu)/8760 \times i_m A_{c(1)}$$

On combining the equations used to determine $A_{c(1)}$, T, and i_c, the value of T can be expressed in following terms.

$$T = (wCu\alpha\gamma)/[8760(\varnothing_{corr} - \varnothing_c) \times (2\pi r\rho L_{as})]$$

In the above equations, the terms $2\pi r\rho L_{as}$ and R_a/γ are equivalent to value of slope parameter S, and calculating this equivalence is also referred to as the slope parameter method of CP design. It is important to note that the magnitude of the slope parameter differs from that of bare steel by a factor of $1/\gamma$. If we take the mean CP design current density (i_m) to be 7 mA/ft^2

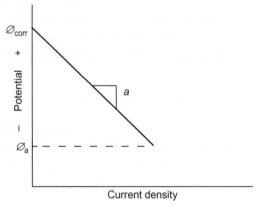

Figure 14.1 Proposed potential versus current density plot.

when the closed circuit cathode potential value($\emptyset_c$) is -80 $V_{Ag/AgCl}$, then the value of polarization resistance (α) will be ≥ 2.0 Ω m^2. Thus, a realistic limit for the coating breakdown factor is 7%. When the ratio of the total pipe surface area to the pipe's bare surface area (γ) is 14.3, the lower limit value of 7% $\alpha\gamma$ is 30 Ω m^2. Initially, this leads to a baseline value for anode spacing (L_{as}) in terms of anode surface area or dimensions using McCoy's formula. More simply put, the anode spacing value is stated in terms of α, γ, R_a, and design cathode potential ($\emptyset_c$). The value of L_{as} can be substituted into the following equation where the design life T is expressed in terms of anode weight (w), anode current capacity (C), anode utilization factor (u), attenuation factor (α), ratio of total surface area to bare surface area (γ), corrosion potential closed circuit ($\emptyset_{corr}$), anode potential ($\emptyset_c$), pipeline dimensions (radius $r\rho$), and anode spacing (L_{as}). Using this equation, the values of w and T can then be optimized. Given that anode dimensions and weight are interrelated and change over time, the changes in L_{as} and α γ results in different values of anode potential and CP design life, they should be accordingly changed to optimize the values of w and T.

SAMPLE CALCULATIONS

A 273-mm-diameter pipeline that has a wall thickness of 8.54 mm is designed for 30 years life. The selected anode weighs 60.8 kg and is 432 mm long. The outré radius of the anode is 178 mm.

The following data about the design are available:

Local sea water (electrolyte) resistivity: 0.80 Ω m

Attenuation resistance (α): 7.5 Ω m^2

Ratio of the total pipeline surface area to the coating breakdown area (γ): 19.89

Anode current capacity: 1700 Ah/kg

Anode utilization factor (u): 0.8

Open circuit anode potential: -1.05 $V_{Ag/AgCl}$

Closed circuit anode potential ($\emptyset_c$): -974 $V_{Ag/AgCl}$

We calculate the resistance of an individual anode as $(R_a) = 0.353$ Ω, using the McCoy formula, in which

$$R_a = (0.315\rho_e)/\sqrt{A_a}$$

The calculation of anode spacing (L_{as}) uses the following equation, discussed above,

$$L_{as} = [(\varnothing_c - \varnothing_a)/\varnothing_{corr} - \varnothing_c] \times [(\alpha\gamma)/(2\pi r \rho R_a)]$$

The above calculation then produces a value of 170,000 mm (170 m). Similarly, the calculation also agrees with the value of 30 years for the design life.

In the above example, if the calculated values were to differ significantly from the required values, then alternative values for any combination of anode weight (w) which is equalized to individual anode resistance (R_a) should be used with the attenuation factor (α), the ratio of the total pipe surface area to the coating breakdown area (γ), and the anode spacing value (L_{as}) to optimize the result.

CHAPTER FIFTEEN

Attenuation Modeling for Offshore Pipelines

INTRODUCTION

In the previous chapters, we discuss the need for attenuation calculations, and we describe the principles that apply to calculations of potential attenuation in cathodically polarized pipelines. In these discussions, we imply that the Morgan-Uhlig equation and boundary-element modeling (BEM) have limited explanatory power, as they respectively exclude anode resistance and metallic path resistance from their calculation approaches. We also note that the previously discussed slope parameter approach to CP design excludes the metallic path resistance as well. To address these limitations, we introduce an approach that incorporates the following four resistance terms in the modeling of CP systems.

1. Anode resistance
2. Coating resistance
3. Polarization resistance
4. Metallic path resistance

Often referred to as first-principles-based, the equations that support the principles of the attenuation model are discussed in this chapter.

EQUATIONS, TERMS, AND EXPRESSIONS

If a metallic pipeline is considered to be an electrode, then its potential at a point z along the pipeline is represented by

$$\emptyset_c(z)$$

This potential can be represented as a charge gradient by using the following equation, where the potential on the pipe side of the double layer is

$U_m(z)$ and the seawater potential just outside the double layer is $U_e(z)$. The resulting value is constant to account for the use of reference potential as $\emptyset_c(z)$ at point z, as discussed above.

$$\emptyset_c(z) = U_m(z) - U_e(z) + K_{ref}$$

In the above equation, the magnitude of the polarization ($E_c(z)$) is the magnitude of the cathodic polarization at any point z on the pipeline. If this equation is combined with the second derivative of the above equation, then we get the following relation;

$$\Delta^2 E_c / \Delta z^2 = \left(\Delta^2 U_m / \Delta z^2 \right) - \left(\Delta^2 U_e / \Delta z^2 \right)$$

The value of cathodic polarization (E_c), the potential of the pipe (U_m), and the seawater potential (U_e) are expressed in terms of their second derivatives, and these second derivatives are substituted into the above equation, resulting in an equation called the finite difference method (FDM). FDM is derived by an iterative and explicit finite difference approach, and many universities have conducted detailed work on these principles. FDM appears as follows.

$$\Delta^2 E_c(z)/\Delta z^2 = \Delta^2 E_c(z)/\Delta z H[1/r_a - 1/z) + E_c(z) \times \left[(2H/z^2) - B \right]$$

$$= 2H \times 1/z^3 \int_z^L E_c(z^*) dz$$

In the above equation, H is a ratio with the pipeline-to-seawater resistivity (ρ_e) and pipe radius (r_p) as the numerator, and the polarization resistance (α) and ratio of the total pipe area to the coating breakdown holidays area (γ) as the denominator. The value of r_a is the radius of the anode that is attached on the pipe at the intervals of $2L$. Similarly, the value of B is the pipeline metallic path resistance $R_m \times 2\pi r_p/\alpha$ (polarization resistance) and γ (ratio of total pipe area to the coating breakdown holidays area).

The FDM, along with the BEM, can be used to conduct verification of the CP design. The above equation can also be used to verify the accuracy of the slope parameter design method discussed in Chapter 14 of this section, where use of the following equations were made to derive the values of anode spacing and the CP design life.

$$L_{as} = [(\emptyset_c - \emptyset_a)/\emptyset_{corr} - \emptyset_c] \times [(\alpha\gamma)/(2\pi r \rho R_a)]$$

And

$$T = (wCu\alpha\gamma)/[8760(\emptyset_{\text{corr}} - \emptyset_{\text{c}}) \times (2\pi r\rho L_{\text{as}})]$$

The results from the two slope parameter equations are compared with the result from the FDM equation and found to be in agreement, supporting the design. A variation of up to about 3% in the values of the anode spacing L_{as} and $\alpha\gamma$ values is generally considered satisfactory. The FDM approach is an iterative method, whereby different parameters can be included in order to reach the most optimized results in the calculation of spacing and cathode potential.

CRITICAL ISSUES WITH THE RETROFITS

Any successful CP design depends primarily on the current density demand of the structure, whether that structure is a jetty, a spar, an floating production storage and off-loading (FPSO), a pipeline end manifold (PLEM), a platform, or a pipeline. The anode size required for the CP system directly varies with the magnitude of the design current density, while the spacing of anodes varies indirectly to the magnitude of the design current density. These facts often lead to the overdesigning of the anode numbers and their close spacing. A CP retrofit is no exception to this practice.

The FDM can be used to plot an attenuation profile over different time increments. This plot can then be used to project the life of the depleting CP system. The gradual depletion of anodes will result in increased anode resistance and more positive potential values over the years. This change in potential is due to the fact that, for bracelet anodes on platforms, a shrinking anode radius leads to a decrease in current output per anode value I_{a} leading to decreased polarization, while, for other anodes, a reduced cross-sectional area has the same effect. This process can help determine the state of the CP system, thus supporting proper retrofit design.

In case of pipelines, retrofit data collection is restricted by the accessibility of the pipeline for retrofitting. The cost of an ROV, divers, and ship adds to the complexities of retrofitting a pipeline. These challenges make the maximization of anode space the guiding principle for the design of pipeline retrofits. However, maximizing anode space is not always very practical, as the bottom of the sea may have geographical features that may limit the application of the maximization principle. The pipeline current density calculated with the average value ($\emptyset_{\text{c}}$ Av) and the far field value ($\emptyset_{\text{c}}$ FF) often provides an agreeable number to use. The anode current output, which is determined using the FDM and slope parameter approaches, is also used. If pipe resistance is not a factor, then the data from the FDM is considered to be more accurate, as compared to the slope parameter method. In the case that pipe resistance is a factor, the modified FDM equation is often used as well.

ANODE BATTERY DESIGN

An anode battery is an array of anodes designed to provide the required current density to the pipeline, with the anodes depleting at the end of the CP design life. An anode battery may also be designed to provide supplemental current to an existing CP system that was poorly designed. Designing an anode battery involves determining the current density demand, the anode resistance, and the effective CP range in terms of the distance from the battery to the pipeline. The CP range depends on the location of the anode battery and the battery's ability to meet the pipeline's required current demand. As a result, these system requirements can lead to either overdesigning the number of anodes in a battery or mislocating the battery on the sea floor. So, anode battery design clearly demands the optimization of both the number and type of anode in a battery, and the battery must be located so that the CP system can extract maximum benefit from the retrofit design.

Therefore, if the objective of the design is to maximize anode spacing, lowering the individual anode resistance value (R_a) should be the goal. As has been demonstrated, an increase in anode resistance is proportionally related to the electrolyte resistance, but it varies inversely with the anode surface area, $\left[R_a \approx \sqrt{(S_a/4\pi)} \right]$. These relationships emphasize that the resistivity is set to a level where the anode battery is less resistive than the electrolyte. The differing resistances then suggest that, if at all possible, the anode battery should be located on the seafloor, because the resistivity of seafloor mud is always greater than the resistivity of the seawater or anode. As we know, anode resistance is also dependent on the anode's surface area, and to reduce the resistance, multiple anodes are used in the battery or array, where they are attached with common electrical connections. Close form numerical modeling for the resistance is often developed for this optimization.

Pipes also receive protective current from the platforms to which they are attached. In this respect evaluating the role of anodes that are considered noninteracting is important, and retrofitting almost always involves determining resistance for anode interactions. Approaches for determining such resistances include Pierson, Hartt, and Sunde's complex equations. Sunde's approach is used for a single anode or anodes equally spaced in a circle. A rectangular arrangement is also used for constructing an anode battery.

Figure 15.1 is an example of a typical retrofit anode arrangement.

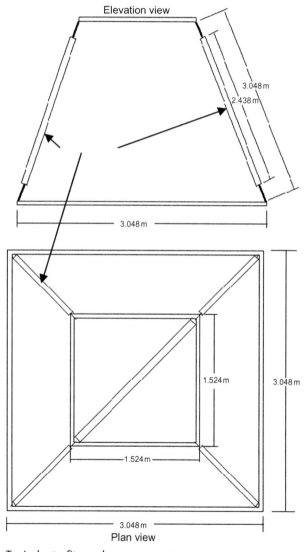

Figure 15.1 Typical retrofit anode arrangement.

The subject of retrofitting has developed significantly over the years, and several papers and articles have been should be referenced in this respect. The following is a list of additional reading that describes retrofitting in detail. Some of these works have been referred to in this section, and others provide more extensive information on various aspects of CP retrofit design.

APPENDIX TO SECTION 5

Symbols and their meanings

A_a	Anode surface area
A_c	Structure (cathode) surface area
$A_c(1)$	Pipe surface area protected by a single anode
C	Anode current capacity
d	Distance from an offset anode to a position on a pipeline
D	Diameter of a circle about which an array of equally spaced anodes are placed
E_a	Magnitude of cathodic polarization at $z=0$
E_b	Magnitude of cathodic polarization at $z=L$
$E_c(z)$	Magnitude of cathodic polarization at z
f_c	Coating breakdown factor (fraction of the external pipe surface that is exposed to coating defects and bare areas)
i_o	Initial cathodic protection design current density
i_m	Mean cathodic protection design current density
i_f	Final cathodic protection design current density
i_c	Structure (cathode) current density demand
$i_c(z)$	Cathodic current density at z along a pipeline
I_a	Current output of an individual anode
$I_m(z)$	Metallic path current in a pipeline at z
k	Polarization resistance of bare metal exposed at coating defects
l	Length of an individual standoff anode
L	Half-spacing between equally spaced anodes
L_{as}	Anode spacing
M	Total anode mass
N	Number of galvanic anode
OF	Distance from an offset anode to a pipeline
r	Anode radius or effective radius
r_a	Equivalent radius of a spherical anode
R_p	Pipeline radius
R	Resistance
$R(N)$	Resistance of an array of N anodes
R_a	Resistance of an individual anode
R_t	Total cathodic protection circuit resistance
R_m	Pipeline metallic path resistance per unit length
S	Distance between equally spaced anodes in a linear array
S	Slope parameter
S_a	Surface area of a cylindrical anode
T	Cathodic protection system design life
u	Anode utilization factor

$U_m(z)$	Potential on the pipe side of the double layer
$U_e(z)$	Electrolyte potential just outside the double layer
v	Volume fraction of the anode that is galvanic metal as opposed to core
w	Weight of an individual galvanic anode
z	Distance along a pipeline from an anode
a'	Attenuation factor
a	Polarization resistance
f_a	Closed circuit anode potential
f_c	Closed circuit cathode potential
f_{corr}	Corrosion potential
$f_c(z)$	Polarized pipe potential at z
$f_c(FF)$	Polarized pipe potential at large z
$f_c(Av)$	Average polarized pipe potential
g	Ratio of total pipe surface area to bare surface area

International Bodies that Address Corrosion, Their Specifications, and an Explanation of Symbols, Constants, and Indices

CHAPTER SIXTEEN

International Bodies that Address Corrosion

INTRODUCTION

Several international bodies address various aspects of corrosion and corrosion control measures. These bodies may adopt different approaches to tackling the same issue, and they may prescribe different standards for acceptance criteria, tests, and testing methods. But, regardless of an organization's approach, the main principles and definitions of corrosion and corrosion prevention and mitigation remain basically the same.

Some national specifications put more emphasis on local elements considered to be more likely to contribute to corrosion. Such elements and local conditions may also find more emphasis in local standards. This chapter focuses on the specifications of international bodies, however, as those guidelines are more or less universally accepted as code or may be the basis for local codes regarding the engineering, design, prevention, and control issues involved with corrosion control. As it introduces each body, this chapter also offers examples of the subjects addressed by that body, but, in fact, the actual list is exhaustive for each group. From time to time, the specifications generated by the bodies are amended, merged, removed, or changed, so the most recently updated version of their specifications must always be used and referenced.

NATIONAL ASSOCIATION OF CORROSION ENGINEERS INTERNATIONAL

The National Association of Corrosion Engineers International (NACE International) was established in 1943 by 11 corrosion engineers from the pipeline industry. The founding engineers were originally part of a regional group formed in the 1930s when the study of cathodic protection (CP) was introduced. Since then, NACE International has become the global leader in developing corrosion prevention and control standards, certification, and education. The members of NACE International still include engineers, as well as numerous other professionals working in a range of areas related to corrosion control.

NACE International, The Corrosion Society, serves nearly 30,000 members in 116 countries, and it is recognized globally as the premier authority for corrosion control solutions. The organization offers technical training and certification programs, conferences, industry standards, reports, publications, technical journals, government relations activities, and more. NACE International is headquartered in Houston, Texas, with offices in San Diego, Kuala Lumpur, and Shanghai. NACE International issues documents and publications relevant to various aspects of corrosion control methods. Not all of those documents can be listed or referenced, but some typical NACE specifications and publications used in CP and coating are listed below.

NACE Standard RP 0169	Control of External Corrosion on Underground or Submerged Metallic Piping Systems

This standard practice (SP) presents procedures and practices for achieving effective control of external corrosion on buried or submerged metallic piping systems. These recommendations are also applicable to many other buried or submerged metallic structures, and they are intended for use by corrosion control personnel concerned with the corrosion of buried or submerged piping systems, including pipes for oil, gas, and water, as well as similar structures. This standard describes the use of electrically insulating coatings, electrical isolation, and CP as external corrosion control methods. It contains specific provisions for the application of CP to existing bare metal and coated piping systems, as well as new structures. Also included are procedures for the control of interference currents on pipelines. For accurate and correct application of this standard, the standard must be used in its entirety. Using or citing only specific paragraphs or sections can lead to misinterpretation and misapplication of the recommendations and practices contained in this standard. This standard does not designate practices for every specific situation because of the complexity of conditions to which buried or submerged piping systems are exposed.

NACE Standard RP 0176	Corrosion of Fixed Offshore Structures Associated with Petroleum Production

This NACE standard provides guidelines for establishing minimum requirements for the control of corrosion on fixed offshore steel structures associated with petroleum production, and on the external portions of associated oil- and gas-handling equipment. Fixed structures include platforms, tension leg platforms, and subsea templates. The guidelines for the control of corrosion on temporarily moored mobile vessels used in petroleum production are not covered by this standard. This standard, divides fixed structures into three zones requiring different approaches to corrosion control. These zones are listed below as:

The submerged zone

The splash zone

The atmospheric zone

As we have discussed in earlier sections, the submerged zone requires sacrificial anodic CP, and this standard addresses that part of the structure. The corrosion control of atmospheric and splash zones are addressed differently, and they are not part of this standard.

NACE RP 0675	Control of External Corrosion on Offshore Steel Pipelines

This standard is withdrawn by NACE and a new standard is planned. However, it is referenced here because of its use in the past and references.

NACE RP 0387	Metallurgical and Inspection Requirements for Cast Sacrificial Anodes for Offshore Applications

This recommended practice (RP) defines minimum physical quality and inspection standards for cast sacrificial anodes for offshore applications. The two main objectives of the standard are listed below.

• To standardize an industry-wide practice that can be used by consultants, manufacturers, and users to define the physical requirements of anodes
• To be specific enough to assist the inspection authority in its task of confirming that anodes comply with the physical requirements

NACE RP 0287	Field Measurement of Surface Profile of Abrasive Blast Cleaned Steel Surface Using a Replica Tape

This standard describes a procedure for the on-site measurement of the surface profile of abrasive blast-cleaned steel surfaces that have a surface profile ranging from 38 to 114 µm (1.5 and 4.5 mils). The procedure correlates with the measurements obtained by the defined laboratory procedure on nonrusted panels prepared in accordance with the NACE No. 1/SSPC-SP 5, NACE No. 2/SSPC-SP 10, or NACE No. 3/SSPC-SP 6. The specification also gives suggestions regarding the implementation and use of this procedure.

NACE RP 0188   Discontinuity (Holiday) Testing of Protective Coatings

---

In 2006, this RP was converted to SP, and it is now available as NACE SP 0188. The specification provides a procedure for the electrical detection of minute discontinuities called "holidays" in coating systems that are liquid-applied to conductive substrates other than pipelines. Procedures are also described for determining discontinuities using two types of test equipment:
1. Low-voltage wet sponge
2. High-voltage spark testers

NACE TM0 190   Impressed Current Laboratory Testing of Aluminum Alloy

---

This standard details the quality assurance procedure for determining the potential and current capacity characteristics under laboratory conditions for aluminum alloy anodes used for CP. The procedure screens various heats or lots of anodes to determine performance consistency on a regular basis from lot to lot. One method for anode potential evaluation and two methods (mass loss and hydrogen evolution) for current capacity evaluations are described. Performance criteria and sampling frequency are left to the discretion of the users of the standard. Keywords: anodes, CP, and testing.

NACE Publication   Design of Galvanic Anode Cathodic Protection Systems for
7L198   Offshore Structures

This publication is similar to several published by NACE on specific subjects. It describes a new design method based on first principles derivations, summarizes laboratory and field experimental data related to the new design approach, gives examples of how existing design criteria are incorporated into the new design equation, and presents two example designs using the new equation. The new design approach allows for the more precise design of CP systems, particularly in deep water or new geographic areas.

NACE and SSPC (The Coating Society: SSPC.Org) have issued joint standards for surface preparation in advance of coating and painting applications, and these standards contain the numbers described below. As the names indicate, the level of cleanliness is described by numbering systems. For example, the NACE No. 1, which is equal to SSPC-SP 5, is described as white metal blast cleaning.

NACE No. 1, SSPC-SP 5   White Metal Blast Cleaning

This joint standard covers the requirements for white metal blast cleaning of unpainted or painted steel surfaces through the use of abrasives. These requirements include the end condition of the surface and materials and procedures necessary to achieve and verify the end condition. A white-metal-blast-cleaned surface, when viewed without magnification, shall be free of all visible oil, grease, dust, dirt, mill scale, rust, coating, oxides, corrosion products, and other foreign matter.

NACE 2/ SSPC-SP 10   Near-White Metal Blast Cleaning

This joint standard covers the requirements for near-white blast cleaning of unpainted or painted steel surfaces through the use of abrasives. These requirements define the end condition of the surface and materials and the procedures necessary to achieve and verify the end condition. A near-white-metal-blast-cleaned surface, when viewed without magnification, shall be free of all visible oil, grease, dust, dirt, mill scale, rust, coating, oxides, corrosion products, and other foreign matter, except for staining, as noted. Random staining shall be limited to no more than 5% of each unit area of surface, as defined, and may consist of light shadows, slight streaks, or minor discolorations caused by stains of rust, mill scale, or previously applied coating.

NACE 3/ SSPC-SP 6   Commercial Blast Cleaning

This joint standard covers the requirements for commercial blast cleaning of unpainted or painted steel surfaces through the use of abrasives. These requirements include the end condition of the surface and materials and procedures necessary to achieve and verify the end condition. A commercial-blast-cleaned surface, when viewed without magnification, shall be free of all visible oil, grease, dust, dirt, mill scale, rust, coating, oxides, corrosion products, and other foreign matter, except for staining, as noted. Random staining shall be limited to no more than 33% of each unit area of surface, as defined, and may consist of light shadows, slight streaks, or minor discolorations caused by stains of rust, mill scale, or previously applied coating.

NACE 4 /SSPC-SP 7	Brush-Off Blast Cleaning

This joint standard covers the requirements for brush-off blast cleaning of unpainted or painted steel surfaces by the use of abrasives. These requirements include the end condition of the surface and materials and procedures necessary to achieve and verify the end condition. A brush-off-blast-cleaned surface, when viewed without magnification, shall be free of all visible oil, grease, dirt, dust, loose mill scale, loose rust, and loose coating. Tightly adherent mill scale, rust, and coating may remain on the surface. Mill scale, rust, and coating are considered tightly adherent if they cannot be removed by lifting with a dull putty knife after abrasive blast cleaning has been performed.

INTERNATIONAL ORGANIZATION FOR STANDARDIZATION

The International Organization for Standardization (ISO) started with a meeting of delegates from 25 countries in 1946, at the Institute of Civil Engineers in London. Theses delegates agreed to create a new international organization "to facilitate the international coordination and unification of industrial standards." In February 1947, the new organization, ISO, officially began operations. The name "International Organization for Standardization" should logically have acronym IOS. However, the word order would not be correct in other languages. For example, in French, the name would be Organisation internationale de normalization, leading to the acronym OIN. So, to avoid such conflicts, the originators of the organization agreed on a more acceptable acronym, ISO, which is derived from the Greek word isos, meaning equal. The acronym's connection to isos reflects that, in whichever country and whatever the language, the short name is always ISO.

Since its founding, ISO has added several members, correspondents, and participants, and as of this writing, ISO has members from 164 countries and functions through 3368 technical bodies that participate in the standard development process. The office of ISO is called the Central Secretariat, and it is located in Geneva, Switzerland. ISO is possibly the world's largest organization engaged in developing voluntary international standards, and it forms a network of national standards bodies. These national standards bodies make up the ISO membership and they represent ISO in their country. The membership is categorized in three levels with different level of access and influence over the system. In ISO's opinion this categorization helps to include all of its members, while also recognizing the different needs and capacities of each national standards body. The three levels and their influences are described as below.

- Full members (or member bodies) *influence* ISO standards development and strategy by participating and voting in ISO technical and policy meetings. Full members sell and adopt ISO International Standards nationally.
- Correspondent members *observe* the development of ISO standards and strategy by attending ISO technical and policy meetings as nonparticipants. Correspondent members can sell and adopt ISO International Standards nationally.
- Subscriber members *keep up to date* on ISO's work but do not attend technical and policy meetings. They do not sell or adopt ISO International Standards nationally.

International Standards provide state-of-the-art specifications for products, services, and good practice, helping to make industry more efficient and effective. There are ~19,500 different standards developed by ISO. These standards are developed through global consensus, and as a result, they are acceptable internationally for various products ranging from food safety to computers. Standards are also developed with the intent of universal application to help break down barriers to international trade. Typical ISO specifications applicable to corrosion prevention and CP follow.

ISO 15589-2	Petroleum and Natural Gas Industries – Cathodic Protection of Pipeline Transportation Systems – Offshore Pipelines
ISO 8501-1	Preparation of Steel Substrate Before Application of Paints and Related Products – Visual Assessment of Surface Cleanliness
ISO 1461	Hot-Dip Galvanized Coating on Fabricated Iron and Steel Articles – Specification and Test Methods
ISO 8044	Corrosion of Metals and Alloys – Basic Terms and Definitions
ISO 21809-1	Petroleum and Natural Gas Industries – External Coating for Buried or Submerged Pipelines Used in Pipeline Transportation Systems (Polyolefin Coatings – 3 Layers of PE and 3 Layers of PP)
ISO 21809-2	Petroleum and Natural Gas Industries – External Coating for Buried or Submerged Pipelines Used in Pipeline Transportation Systems (Fusion Bonded Epoxy Coatings)

ISO 21809-3	Petroleum and Natural Gas Industries – External Coating for Buried or Submerged Pipelines Used in Pipeline Transportation Systems (Field Joint Coatings)
ISO 21809-4	Petroleum and Natural Gas Industries – External Coating for Buried or Submerged Pipelines Used in Pipeline Transportation Systems (Polyethylene Coatings – 2 Layers of PE)
ISO 21809-5	Petroleum and Natural Gas Industries – External Coating for Buried or Submerged Pipelines Used in Pipeline Transportation Systems (External Concrete Coatings)

▷ DET NORSKE VERITAS

Det Norske Veritas (DNV) literally translates to "the Norwegian truth," and it is the name of an independent foundation charged with safeguarding life, property, and the environment. DNV was founded in Norway in 1864 for the purpose of inspecting and evaluating the technical condition of Norwegian merchant vessels. Since then, DNV has concentrated on identifying and assessing risk in different fields, while advising corporations on ways to manage that risk. Such DNV efforts include:

- Classification of ships
- Certification of an automotive company's management system
- Advice how to best maintain an aging oil platform

DNV mainly focuses on the safety and responsible improvement of business, and it uses a unique risk management approach to offer innovative services that meet customers' needs across industries and countries. As a result, DNV enjoys a singular position as a trusted partner in the improvement of quality, safety, and efficiency in high-risk global industries. DNV specifications provide an excellent reference and guide for meeting the basic requirements of CP system design. Typical relevant DNV specifications are listed below.

DNV RP-F103	Cathodic Protection of Submarine Pipelines by Galvanic Anodes
DNV RP-B401	Cathodic Protection Design
DNV RP-F106	Factory Applied External Pipeline Coatings for Corrosion Control
DNV RP-F102	Pipeline Field Joint Coating and Field Repair of Linepipe External Coating

NORSOK

The Norsk Sokkels Konkuranseposisjon (NORSOK) standards are developed by the Norwegian petroleum industry to ensure adequate safety, value-adding, and cost effectiveness for industry developments and operations. Furthermore, the NORSOK standards are intended to replace oil company specifications to the greatest extent possible, while serving as references in the authority's regulations.

NORSOK M 501	Standard for Surface Preparation and Protective Coating
NORSOK M 503	Cathodic Protection, (This Specification Addresses the Cathodic Protection of Submerged Installations and Seawater Containing Compartments and Manufacturing and Installation of Sacrificial Anodes)

AMERICAN SOCIETY FOR TESTING MATERIALS

The American Society for Testing Materials (ASTM) offers specifications that complement most construction specifications, and these specifications and codes address several material and testing procedures and requirements, while offering related guidance. The ASTM specifications (www.astm.org) are organized according to material type, and the letters prefixed to the specification number are indicative of the material type. For example, the letter A is for all ferrous materials; the letter B is for all nonferrous materials; and the letter C is for cementations, ceramic, concrete, and masonry. The letter D is used to indicate specifications-related miscellaneous material, such as chemicals, polymers, paints, coatings, and their test methods, and similarly, the letter E is used to denote specifications that address miscellaneous subjects, including subjects related to the examination and testing of material. The following is a short list of some groups included in the specifications.

ASTM G 8	Test Method for Cathodic Disbonding of Pipeline Coating
ASTM D 1141	Specification for Substitute Ocean Seawater
ASTM D 4060	Standard Guide to Standard Test Methods for Unsintered Polytetrafluoroethylene (PTFE) Extruded Film or Tape

ASTM D 2583	Standard Practice for Fusion of Poly(Vinyl Chloride) (PVC) Compounds Using a Torque Rheometer
ASTM D 185	Test Methods for Coarse Particles in Pigments, Pastes, and Paints
ASTM D 1640	Test Methods for Drying, Curing, or Film Formation of Organic Coating at Room Temperature
ASTM G 6	Test Method for Abrasion Resistance of Pipeline Coatings
ASTM G 9	Standard Test Method for Water Penetration into Pipeline Coatings
ASTM D 10	Standard Test Method for Specific Bendability of Pipeline Coatings
ASTM G 14	Standard Test Method for Impact Resistance of Pipeline Coatings (Falling Weight Test)
ASTM D 2240	Standard Test Method for rubber Propery-Durometer Hardness
ASTM G 11	Test Method for Effects of Outdoor Weathering on Pipeline Coatings
ASTM G 12	Standard Test Method for Nondestructive Measurement of Film Thickness of Pipeline Coating on Steel
ASTM D 792	Standard Test Methods for Density and Specific Gravity (Relative Density) of Plastics by Displacement
ASTM D 1505	Standard Test Method for Density of Plastics by the Density-Gradient Technique
ASTM D 1693	Standard Test Method for Environmental Stress-Cracking of Ethylene Plastics
ASTM D 4138	Standard Practice for Measurement of Dry Film Thickness of Protective Coating Systems by Destructive, Cross-Section Means
ASTM D 4940	Standard Method for Conductimetric Analysis of Water Soluble Ionic Contamination of Blasting Abrasives

CANADIAN STANDARDS ASSOCIATION

The Canadian Standards Association (CSA) is an organization that develops industrial standards spanning over 57 different industrial areas. This not-for-profit organization has representatives from industry, government, and consumer groups and publishes standards CSA is accredited by the Standards Council of Canada, which is a crown corporation that develops and issues standards to promote standardization and efficiency in industrial production processes in Canada. The CSA standards are available in print and electronic form.

Starting as the Canadian Engineering Standards Association (CESA) in the early 1900s, the CSA was created as a result of Sir John Kennedy's efforts to emphasize the need for such an organization. Initially, the organization addressed the needs of industries involved in the production of aircraft parts, the design and construction of bridges, the construction of buildings, and electrical work, among other endeavors. The first standards issued by CESA were for steel railway bridges, in 1920. CESA was renamed the CSA in 1944, and its certification mark was introduced in 1946. The organization now operates in 57 different areas of specialization, including climate change, business management, and safety and performance standards for electrical and electronic equipment, industrial equipment, boilers and pressure vessels, compressed gas-handling appliances, environmental protection, and construction materials. The following list presents some of the widely adopted CSA standards that relate to corrosion protection by way of coating quality control.

CSA Z 245.20	External Fusion Bond Epoxy Coating for Steel Pipe

The CSA Z 245.20 specification covers the qualification, application, inspection, testing, handling, and storage of coating materials. This specification is applicable to fusion-bonded epoxy (FBE) coating applied in the plant. The pipes coated with FBE are intended for use in submerged offshore applications or for pipelines buried underground. The specification covers different types of coating systems and classifies them with alpha–numeric identifiers, as listed below.

1A: Single-layer FBE with glass transition temperature of $\leq 110\,^{\circ}C$
1B: Single-layer FBE with glass transition temperature of $\geq 110\,^{\circ}C$
2A: Two-layer FBE with a corrosion coating and protective overcoat
2B: Two-layer FBE with a corrosion coating and an abrasion-resistant overcoat
2C: Two-layer FBE with a corrosion coating and an antislip overcoat
3: Three-layer FBE with an antislip overcoat applied over a corrosion coating and protective overcoat

CAS Z 245.21	External Polyethylene Coating for Steel Pipe

The CSA Z 245.20 specification covers the qualification, application, inspection, testing, handling, and storage of coating materials. This specification is applicable to the external coating of pipes with polyethylene. The pipes with this coating system are primarily intended for use in submerged offshore applications.

There are several other bodies that issue specifications, RPs, and test methods, and their guidelines are often widely referenced and used internationally. Among them are documents from Deutsches Institut für Normung, which is commonly referred as DIN, and Euro Norms (EN). As the previous lists suggest, all these specifications, RPs, and test methods address a specific area of the subject, and they are to be used accordingly. Use of the most recent issue of a specification document is always recommended, unless the situation demands the use of a previously issued version.

CHAPTER SEVENTEEN

Commonly Used Constants, Quantities, and Symbols

This chapter presents the indices, constants, quantities, and symbols most commonly used in equations related to corrosion prevention, and specifically, the design and interpretation of cathodic protection. As described in previous chapters, these symbols and expressions apply to the cathodic protection of offshore structures, but they are also used universally in the field. The common indices are listed below.

Electrical Quantities Y

Y' = Length-related quantity (Y-load)

Y_x = Y at the point coordinates x (e.g., r, 1, 0, ∞)

Y_X = Y for a definite electrode or object X

Chemical and Thermodynamic Quantities Y

Y° = Standard conditions

Y^* = Conditions for thermodynamic equilibrium

Y_i = Quantity of component X_i

Electrochemical Quantities Y

$Y_{a,c}$ = Quantities of the anodic (a) or cathodic (c) region as well as the relevant total currents

$Y_{A,C}$ = Quantity of anodic (A) or cathodic (C) partial reaction

Y_e = Quantity in cell formation

General Symbols Used

EP = Epoxy resin

FI = Failure current

FU = Failure voltage

e^- = Electron

PN = Nominal pressure

HV = Vickers hardness

HV_{dc} = High-voltage dc transmission

IR = Ohmic voltage drop

Me = Metal

Ox = Oxidizing agent

PE = Polyethylene

PUR = Polyurethane
PVC = Polyvinylchloride
Red = Reducing agent
X_i = Symbol for material i

SYMBOLS

The following table lists the symbols used in the text, as well as the formulas that employ the symbols. These symbols are not limited to this text alone but can be found in various other texts, papers, and presentations.

Symbol	Meaning	S.I. Units
a	Distance, length	cm, mm, m
b	Distance, length (second)	cm, mm, m
$b_{+/-}$	Logarithmic relationship between current and applied voltage, as $\eta = a + b \log i$, where η is overvoltage, i is the current, and a and b are constants. This is also called Tafel slope (log)	mV
B	Mobility	$cm^2 \, mol \, J^{-1} \, s^{-1}$
B_0	Soil aggressiveness, total rating number	
B_1	Soil aggressiveness, total rating number	
B_e	Soil aggressiveness, total rating number	
$c_{(X_i)}$	Concentration of material X_i	$mol \, cm^{-3}$
C	Capacity or constant	
C_D	Double-layer capacity of an electrode	$\mu F \, cm^{-2}$
d	Diameter	cm, mm
D_i	Diffusion constant of material X_i	$cm^2 \, s^{-1}$
E	Electric field strength	$V \, cm^{-1}$
f	Frequency	$Hz = s^{-1}$
f_a	Conversion factor	$mm \, a^{-1} / (mA \, cm^{-2})$
f_b	Conversion factor	$gm^{-2} h^{-1} / (mA \, cm^{-2})$
f_c	Conversion factor	$mm \, a^{-1} / (gm^{-2} \, h^{-1})$
f_v	Conversion factor	$Lm^{-2} h^{-1} / (mA \, cm^{-2})$
F	Force	N
F	Interference factor	
F	Faraday constant $= 96485 \, A \, s \, mol^{-1} = 26.8 \, A \, h \, mol^{-1}$	
g	Limiting current density	$A \, m^{-2}$
G	Limiting current	A

G	Leakage	$S = \Omega^{-1}$
G'	Leakage per unit length	$S\,m^{-1}$, $S\,km^{-1}$
ΔG	Free enthalpy of formation	$J\,mol^{-1}$
h	Height or soil covering	cm, m
i	Run number	
I	Current	A
I_s	Required protection current	A
I'	Current supply	$A\,km^{-1}$ or $A\,mile^{-1}$
j_H	H-permeation rate	$L\,cm^{-2}/min$
J_i	Transport rate of material X_i	$mol\,cm^{-2}\,s^{-1}$
J	Current density	$A\,m^{-2}$ or $mA\,cm^{-2}$
J_{max}	Maximum current density of sacrificial anode	$A\,m^{-2}$ or $mA\,cm^{-2}$
J_P	Passivation current density	$A\,m^{-2}$ or $mA\,cm^{-2}$
J_s	Protection current density	$A\,m^{-2}$ or $mAcm^{-2}$
J_0	Exchange current density	$A\,m^{-2}$ or $mA\,cm^{-2}$
k	Polarization parameters	cm, m
K	Stress intensity	$N\,mm^{-1.5}$
K	Equilibrium constant	$1\,(mol\,L^{-1})^{(\Sigma n_j)}$
K_{sx}	Acid capacity up to $pH = x$	$mol\,L^{-1}$
K_{Bx}	Base capacity up $pH = x$	$mol\,L^{-1}$
K_W	Ionization constant for water ($10^{-14}\,mol^{-2}\,L^{-2}$ at 25 °C)	
K_w	Reaction constant in oxygen corrosion	mm
l	Length, as in distance	cm, m, mm
l_i	Ion mobility of material X_i	$S\,cm^2\,mol^{-1}$
l_k	Nominal length	m, km, mile
L	Protection range in length units	m, cm, mm
L	Inductivity	$H = \Omega\,s$
L_{Gr}	Limiting length in length units	m, cm, mm
m	Mass	G, kg, lb
m'	Pipe mass per unit length	kg/m
M	Atomic, molecular weight	$g\,mol^{-1}$
M'	Mutual inductivity per unit length	
n	Number or number of cycles	
n'	Number per unit length	
n_i	Stoichiometric coefficient charge number of material X_i	
N	Defect as coating holiday density	m^{-2}
N	Reciprocal slope of $\ln|J|$-U-curves	mV

p	Pressure, gas pressure	bar, psi
$P_{(X_i)}$	Partial pressure of component X_i	bar, psi
Q	Electric charge	A s, A h
Q'	Current constant of sacrificial anode per unit mass	A h kg^{-1} or A h lb^{-1}
Q''	Current constant of sacrificial anode per unit volume	A h dm^{-3}
r	Radius as distance	In, cm, m
r	Reduction factor	
R_p	Specific polarization resistance	Ω m^2
r_u	Specific coating resistance	Ω m^2
R	Electrical resistance	Ω
R	Gas constant $= 8.31$ J mol^{-1} K^{-1}	
R'	Resistance per unit length	Ω m^{-1}, Ω ft^{-1}
R_m	Ultimate tensile strength	N mm^2, psi
R_p	Polarization resistance	Ω
$R_{p0.2}$	0.2% proof stress	N mm^2, psi
R_u	Coating resistance	Ω
s	Distance as in thickness, reduction in thickness	In, mm, cm
S	Surface, cross section	In^2, m^2
t	Depth	In, cm, m
T	Temperature	°C, °F, K
u_i	Electrochemical mobility of substance Xi	V^{-1}cm^2 s^{-1}
U	Potential measured in voltage	V
U_{off}	Off potential	V
U''_B	Potential difference between reference electrode parallel over the pipeline	mV, V
$U_{\perp B}, \Delta U_x$	Potential difference between reference electrode perpendicular to the pipeline at distant x	mV, V
$U_{Cu-CuSO_4}$	Potential measured against Cu-CuSO$_4$ reference electrode	mV, V
U_{on}	Potential measured in on-position	V
U_H	Potential measured against standard hydrogen electrode	mV, V
U_{IR}	Ohmic voltage drop	V
$U_{IR-free}$	IR-free potential	V
U_R	Rest potential	V
U_s	Protection potential	V
U_T	Driving voltage	V
U_{over}	Reverse switching potential	V
U_0	Open circuit potential (EMF)	V
v	Weight loss per unit area time	G m^{-2} h^{-1}
v_{int}	Mean value of v	G m^{-2}
V	Volume	In^3, cm^3, mm^3

V	Atomic, molecular volume	$m^3\,mol^{-1}$, $L\,mol^{-1}$
w, w_{int}	Thickness reduction rate	$In\,a^{-1}$, $mm\,a^{-1}$
w	Degree of effectiveness	%
w	Number of winding	
w_i	Velocity of material X_i	$In\,s^{-1}$, $cm\,s^{-1}$
x	Position coordinate	in., mile; m, km,
Y'	Admittance per unit area	$S\,mile^{-1}$, $S\,km^{-1}$
Y_s	Yield point	psi, $N\,mm^{-1}$
Z	Impedance	Ω
Z_i	Rating number of soil aggressiveness	
z_i	Charge number of material X_i	

GREEK SYMBOLS

Symbol	Meaning	Units
α	Symmetry factor	
α	DC path constant	$Mile^{-1}$, km^{-1}
$\beta_{+/-}$	Tafel slope	mV
γ	Transfer coefficient	$Mile^{-1}$, km^{-1}
δ	Diffusion layer thickness	in., mm, cm
ε, ε_r	Relative dielectric constant	
ε_o	Electric field constant $(8.85 \times 10^{-14}\,F\,cm^{-1})$	
η	Over voltage, polarization	mV, V
η_Ω	Ohmic voltage drop, resistance polarization	mV, V
x	Specific conductance	$S\,cm^{-1}$, $S\,in^{-1}$
μ_i	Partial polar free enthalpy of material X_i	$J\,mol^{-1}$
μ_o	Magnetic field constant (1.26×10^{-8})	$H\,cm^{-1}$, $H\,in^{-1}$
υ	Relative number of cycles	
ρ	Resistivity	$\Omega\,cm$, $\Omega\,in$
ρ_{st}	Specific resistance of steel (ca. $1.7 \times 10^{-6}\Omega\,m$)	$\Omega\,cm$, $\Omega\,in$
ρ_s	Density	$G\,cm^{-3}$, $lb\,in^{-3}$
σ	Tensile strength	$N\,cm^{-2}$, psi
τ	Time constant	s
φ	Electrical potential	V
φ	Phase angle	
ω	Cyclic frequency	s^{-1}

AMERICAN WIRE SIZES

In cathodic protection systems, various electrical connections are made using electrical wires and cable. These wires and cables are of various sizes and cross-sections, based on variations in the constituent materials' resistance to the flow of current. Designers must ascertain these values for proper evaluation of cathodic current flow and final polarization of the structure.

The following table lists wire sizes in American Wire Gauge (AWG) numbers and equivalent metric sizes (cross-section area) in square millimeters. Each wire resistance is given for a 1000 ft length.

AWG Number (#)	Metric Size (mm^2)	Nominal Weight (lb 1000 ft^{-1})	Resistance (Ω 1000 ft^{-1})	DC Current Rating (A)
12		20.16	1.65	20
	4	24.37	1.41	24
10		32.06	1.02	30
	6	36.56	0.939	31
8		50.97	0.64	40
	10	60.93	0.588	42
6		81.05	0.41	55
	16	97.49	0.351	56
5		102.2	0.284	63
4		128.9	0.259	70
	25	152.3	0.222	73
2		204.9	0.162	90
	35	213.3	0.16	92
1		258.4	0.129	110
	50	304.07	0.118	130
1/0		325.8	0.102	135
2/0		410.9	0.079	165
4/0		653.3	0.05	230

APPENDIX

Reading List and Bibliography for Cathodic Protection, High Efficiency Coating, and Corrosion Control

INTRODUCTION

In the bibliography, the reference materials pertaining to corrosion, cathodic protection, coatings, and surface preparations are intermixed. This intermixing results from the fact that we cannot discuss any one of these subjects in exclusion of the others. The overlapping nature of the subjects then complicates any effort to separate the references in a bibliography, without either duplicating or excluding a specific section or chapter. As a result, this chapter presents a combined reference list.

The following books and publications, including specifications, are intended to serve as additional reading and reference materials related to the subject of this book. The specifications of various bodies need to be verified, given possible changes in the most recently published versions. However, older and even withdrawn specifications are frequently referenced for projects based on specific engineering needs. Care must be taken to use the version of the specification document that best addresses the task at hand.

BIBLIOGRAPHY AND ADDITIONAL RECOMMENDED READING

1. Corrosion and Corrosion Control – An Introduction to Corrosion Science and Engineering – By Herbert H. Uhlig and R. Winston Revie, A John Wiley Publication, John Wiley & Sons. ISBN: 0-471-07818-2.
2. Handbook of Corrosion Protection – Theory and Practice of Electrochemical Protection Process – By W. von Baeckmann, W. Schwenk, and W. Pinz. Gulf Professional Publishing. ISBN: 0-88415-056-9.
3. Peabody's Control of Pipeline Corrosion – By A.W. Peabody, NACE International, The Corrosion Society. ISBN: 1-57590-092-0.

4. Localized Corrosion – Cause of Metal Failure, ASTM Special Technical Publication 516, By American Society for Testing Materials. 04-51600-27.
5. Corrosion Engineering – By Mars G. Fontana, McGraw Hill Custom Publications. ISBN: 0-07-293973-7.
6. Fundamentals of Designing for Corrosion Control – By R. James Landrum, Published by NACE. ISBN: 0-915567-34-2.
7. J. S. Mandke, "Corrosion Causes Most Pipeline Failures in the Gulf of Mexico," Oil and Gas Journal, October 29, 1990, p. 40.
8. "Improving the Safety of Marine Pipelines," Committee on the Safety of Marine Pipelines, Marine Board, National Research Council, Washington, D.C., 1994.
9. Minerals Management Service Data Base, Gulf of Mexico Region, New Orleans, LA.
10. C. Weldon and D. Kroon, "Corrosion Control Survey Methods for Offshore Pipelines," Proceedings. International Workshop on Offshore Pipeline Safety, New Orleans, Dec. 4-6, 1991, p. 196.
11. "International Workshop on Corrosion Control for Marine Structures," Eds: G.R. Edwards, W. Hanzalek, S. Liu, D.L. Olson, and C. Smith, American Bureau of Shipping, Houston, 2000.
12. Cathodic Protection of Offshore structures, Galvanic and Impressed Current Systems; J.A. Burgbacher; NACE Material Performance, April 1968, pp. 26-29.
13. Cathodic Protection and Coating Design and Performance of a Corrosion Protection System for a North Sea Jacket. Rolf E Lye, NACE CORROSION Paper number 282 April 1969 New Orleans Convention Center.
14. Corrosion and Its Control, H. Van Droffelaar and J.T.N. Atkinson, NACE International Publication. ISBN: 1-877914-71-1.
15. Corrosion Engineering, Mars G. Fontana, Norbert D. Greene, The Southeast Book Publication, McGraw-Hill Book Company. ISBN: 0-07-021461-1.
16. Pipeline Integrity Handbook: Risk Management and Evaluation. By Ramesh Singh Elsevier, Gulf Professional Publication. ISBN: 978-0-12-387825-0.
17. Corrosion Data Survey, NACE Publication. ISBN: 0-915567-07-5.
18. Corrosion Inhibitors Edited by C.C Nathan, NACE Press; Library of Congress Catalog Number 73-85564.

19. NACE: Use of Initial Current Density in Cathodic Protection Design. Sheldon Evans CORROSION 87, Paper number 58. San Francisco California.

20. NACE: Optimizing Retrofitting of Offshore Cathodic Protection Systems, James A. Brandt. CORROSION 93, Paper number 518.

21. NACE Publication 6A 195, Introduction to Thick-Film Polyurethanes, March 1996, TS Group T-6A-67 Report.

22. NACE/AWS/SSPC Joint standard. NACE No.12; Specification for the Application of Thermal Spray Coating (Metallizing) of Aluminum, Zinc, and Their Alloys and Composites for Corrosion Protection of Steel.

23. NACE International Glossary of Corrosion Related Terms, 2009.

24. NACE and ASTM Combined Document; NACE ASTM G 193, Standard Terminology and Acronyms Relating to Corrosion.

25. NACE International Technical Committee Publication Manual. March 2010, Approved by Technical and Research Committee.

SPECIFICATIONS

The following list of specifications introduces the sources on which corrosion prevention systems are based. The most recent versions of the specifications are used. However, there are cases in which older versions are specifically used and referenced, possibly because the project is based on an older version or requires the use of older data.

1. Recommended Practice DNV RP C 203, Fatigue Design of Offshore Steel Structures.

2. Recommended Practice DNV-RP-B401 Cathodic Protection Design.

3. Recommended Practice DNV-RP-F103 Cathodic Protection of Submarine Pipeline by Galvanic Anodes.

4. NACE Standard RP 0169; Recommended Practice: Control of External Corrosion on Underground or Submerged Metallic Piping Systems.

5. NACE RP 0188, Discontinuity (Holiday) Testing of New Protective Coating on Conductive Substrates.

6. NACE Recommended Practice, Liquid Epoxy Coating for External Repairs Rehabilitation and Weld Joints on Buried Steel Pipelines.

7. NACE Standard RP 0176, Recommended Practice, Corrosion Control of Steel Fixed Offshore Structures Associated with Petroleum Production.

8. NACE Standard RP 0178, Fabrication details Surface Finish Requirements and Proper Design Considerations and Vessels to be Lined for Immersion Services.
9. CSA Z 225.20-02 External Fusion Bonded Epoxy Coating for Steel Pipes, Canadian Standard Association. Specification.
10. CSA Z 225.21-02 External Fusion Bonded Epoxy Coating for Steel Pipe/External Polyethylene Coating for Pipe, Canadian Standard Association. Specification.
11. ISO 15589-2, Petroleum and Natural Gas Industries – Cathodic Protection of Pipeline Transportation Systems – Part 2: Offshore Pipelines.
12. ISO 11124-1, Preparation of Steel Substrate Before Application of Paints and Related Products – Specification for Metallic Blast-Cleaning Abrasive, Part 1 through 5.
13. ISO 21809-1 Petroleum and Natural Gas Industries – External Coating for Buried or Submerged Pipelines Used in Pipeline Transportation Systems – Part-1: Polyethylene Coatings (3-layer PE and 3-layer PP).
14. API 5L2, Recommended Practice for Internal Coating of Line Pipe for Non-corrosive Gas Transmission Services. American Petroleum Association Publication.
15. DIN 30678, Polypropylene Coating for Steel Pipes, Deutsche Norm.
16. DIN 30670, Polyethylene Coating for Steel Pipes and Fittings, Deutsche Norm.
17. ASTM D 714, Standard Method for Evaluating Degree of Blistering of Paints.
18. ASTM G 95, Standard Test Method for Cathodic Disbondment Test of Pipeline Coating (Attached Cell Method).
19. ASTM D 3359, Standard Test Method for Measuring Adhesion by Tape Test.
20. ASTM D 4541, Standard Test Method for Pull-off Strength of Coating Using Portable Adhesion.

ADDITIONAL LIST OF READING ON RETROFITTING

1. Clara, Ventura, CA 93001, October 25, 1985.
2. J. S. Mandke, "Corrosion Causes Most Pipeline Failures in the Gulf of Mexico," Oil and Gas Journal, October 29, 1990, p. 40.

3. "Improving the Safety of Marine Pipelines," Committee on the Safety of Marine Pipelines, Marine Board, National Research Council, Washington, D.C., 1994.

4. Minerals Management Service Data Base, Gulf of Mexico Region, New Orleans, LA.

5. C. Weldon and D. Kroon, "Corrosion Control Survey Methods for Offshore Pipelines," Proceedings International Workshop on Offshore Pipeline Safety, New Orleans, Dec. 4-6, 1991, p. 196.

6. "International Workshop on Corrosion Control for Marine Structures," Eds: G.R. Edwards, W. Hanzalek, S. Liu, D.L. Olson, and C. Smith, Am Bureau of Shipping, Houston, 2000.

7. J. Britton, "Continuous Surveys of Cathodic Protection System Performance on Buried Pipelines in the Gulf of Mexico," Paper no. 422 presented at CORROSION/92, Nashville, April 26-30, 1992.

8. "Corrosion Control of Steel-Fixed Offshore Platforms Associated with Petroleum Production," NACE Standard RP 0176, NACE, Houston, 1976.

9. "Cathodic Protection Design," DNV Recommended Practice RP401, Det Norske Veritas Industri Norge AS, 1993.

10. "Corrosion Control of Steel-Fixed Offshore Platforms Associated with Petroleum Production," NACE Standard RP 0176, NACE, Houston, 1976.

11. Wang, W., Hartt, W. H., and Chen, S., Corrosion, vol. 52, 1996, p. 419.

12. W. H. Hartt, Chen, S., and Townley, D. W., Corrosion, vol. 54, 1998, p. 317.

13. Townley, D. W., "Unified Design Equation for Offshore Cathodic Protection," Paper no. 97473 presented at CORROSION/97, March 9-14, 1997, New Orleans.

14. "Design of Galvanic Anode Cathodic Protection Systems for Offshore Structures," NACE International Publication 7L198, NACE International, Houston, TX, 1998.

15. Dwight, H. B., Electrical Engineering, Vol. 55, 1936, p. 1319.

16. Sunde, E. D., Earth Conduction Effects in Transmission Systems, Dover Publications, Inc., New York, 1968.

17. McCoy, J. E., Transactions Institute of Marine Engineers, Vol. 82, 1970, p. 210.

18. Cochran, J. C., "A Correlation of Anode-to-Electrolyte Resistance Equations Used in Cathodic Protection," Paper no. 169 presented at CORROSION/82, March 22-26, 1982, Houston.

19. Strommen, R., Materials Performance, Vol. 24(3), 1985, p. 9.

20. Cochran, J. C., "Additional Anode-to-Electrolyte Resistance Equations Useful in Offshore Cathodic Protection," Paper no. 254 presented at CORROSION/84, April 2-6, 1984, New Orleans.

21. Foster, T., and Moores, V. G., "Cathodic Protection Current Demand of Various Alloys in Sea Water," Paper no. 295 presented at CORRO-SION/86, March 17-21, 1986, Houston.

22. Mollan, R. and Anderson, T. R., "Design of Cathodic Protection Systems," Paper no. 286 presented at CORROSION/86, March 17-21, 1986, Houston.

23. Fischer, K. P., Sydberger, T. and Lye, R., "Field Testing of Deep Water Cathodic Protection on the Norwegian Continental Shelf," Paper no. 67 presented at CORROSION/87, March 9-13, 1987, San Francisco.

24. Fischer, K. P. and Finnegan, J. E., "Cathodic Protection Behavior of Steel in Sea Water and the Protective Properties of the Calcareous Deposits," Paper no. 582 presented at CORROSION/89, April 17-21, 1989, New Orleans.

25. Schrieber, C. F. and Reding, J., "Application Methods for Rapid Polarization of Offshore Structures," Paper no. 381 presented at COR-ROSION/90, April 23-27, 1990, Las Vegas.

26. Burk, J. D., "Dualnode Field Performance Evaluation – Cathodic Protection for Offshore Structures," Paper no. 309 presented at CORRO-SION/91, March 11-14, 1991, Cincinnati.

27. "Pipeline Cathodic Protection – Part 2: Cathodic Protection of Offshore Pipelines," Working Document ISO/TC 67/SC 2 NP 14489, International Standards Organization, May 1, 1999.

28. Morgan, J., Cathodic Protection, Macmillan, New York, 1960, pp. 140-143.

29. Uhlig, H. H. and Revie, R. W., Corrosion and Corrosion Control, Third Edition, John Wiley and Sons, New York, 1985, p. 223.

30. Strommen, R. and Rodland, A, Materials Performance, Vol. 20 No. 10, 1981, p. 7.

31. McCoy, J. E., "Corrosion Control by Cathodic Protection – Theoretical and Design Concepts for Marine Applications," The Institute of Marine Engineers Transactions, Vol. 82, 1970, p. 210.

32. Chapra, C. S. and Canale, R. P., Numerical Methods for Engineers, McGraw-Hill, Second Ed., New York, 1988, pp. 734-737.

33. P. Pierson and W. H. Hartt, "Galvanic Anode Cathodic Polarization of Steel in Sea Water: Part IV – Conductor Arrays on Petroleum Production Platforms," Corrosion, Vol. 55, 1999, p. 686.

34. T. Andersen and A. Misund, "Pipeline Reliability: An Investigation of Pipeline Failure Characteristics and Analysis of Pipeline Failure Rates for Submarine and Cross-Country Pipelines," J. Pet. Technology, April, 1983, p. 709.

35. "Analysis of the MMS Pipeline Leaks Report for the Gulf of Mexico," Texaco USA, 133 W. Santa.

INDEX

Note: Page numbers followed by "*f*" indicate figures, and "*t*" indicate tables.

Printed and bound by CPI Group (UK) Ltd, Croydon, CR0 4YY

08/05/2025

01864900-0001